BKSTS
Dictionary of Image Technology

BKSTS
Dictionary of Image Technology

Focal Press

London and Boston

Focal Press

is an imprint of the Butterworth Group

which has principal offices in

London, Boston, Durban, Singapore, Sydney, Toronto and Wellington

First edition published in 1983 as
BKSTS Dictionary of Audio-visual Terms
Revised and extended 1988

© Butterworth & Co (Publishers) 1988

British Library Cataloguing in Publication Data

BKSTS dictionary of image technology. – Rev. and extended.
  1. Audiovisual equipment. Encyclopaedias
  I. British Kinematograph Sound and Television Society
  II. Happé, L. Bernard, (Louis Bernard)
  621.38′044′0321

  ISBN 0–240–51276–6

Library of Congress Cataloging-in-Publication Data

BKSTS dictionary of image technology.

  Rev. ed. of: BKSTS dictionary of audio-visual terms. 1983
  1. Audio-visual equipment–Dictionaries. I. British Kinematograph Sound and Television Society. II. BKSTS dictionary of audio-visual terms.
III. Title: Dictionary of image technology.
TS2301.A7 B37 1988   621.38′044′0321   88-30221
ISBN 0–240–51276–6

Typeset by Scribe Design, 123 Watling Street, Gillingham, Kent.
Printed and bound by Hartnolls Ltd, Bodmin, Cornwall

Preface to the second edition

The past six years have seen continued development in all the technologies which we took as our original field and in preparing this fully revised edition we felt that its wide scope is better described as a *Dictionary of Image Technology* than by its previous title, *Dictionary of Audio-visual Terms*; in it, we have endeavoured to reflect both current practice and new proposals still being debated internationally.

As Editor I am grateful to all the members of the original committee who have contributed new material and to Stephen Lowe whose glossary in the BKSTS publication *Understanding Video* suggested several additions.

Bernard Happé, *Editor*

May 1988

Preface

As its name indicates, the British Kinematograph Sound and Television Society is concerned with the practice of a wide range of technologies, and in undertaking the preparation of this dictionary of technical terms it was agreed that the work should reflect fully the extensive interests of the Society. We have therefore interpreted the term *audio-visual* in the most generous sense, embracing the preparation and presentation of pictures and sound by film and video as well as by tape-slide, film-strip and multivision.

Many of the boundaries which formerly divided photographic and electronic methods of recording and reproduction have tended to disappear during recent years and the interchange of these media now frequently provides an additional tool in the hands of the creative producer as well as extended facilities for exhibition. Motion picture film and still slides are incorporated in video productions, videotape is transferred to film, and tape–slide shows are converted to both film and videocassettes for alternative methods of distribution. There is thus ample justification for treating these varied disciplines within the same volume on a common basis.

In recognition of current developments, we have also considered it necessary to explain numerous terms in the field of computer practice, which now enters into many techniques of both production and presentation. The actual display of images in computer graphics and computer animation is an obvious example but computer-controlled operations in video and sound production, in animation photography and in motion picture processing are of increasing importance, while complex multi-screen shows involving dozens of projectors and hundreds of individual slides benefit greatly in both programming and presentation from the facilities available through microprocessors.

We hope therefore that by providing brief explanations of a number of such terms we shall help the understanding of these important applications. In these, as throughout the volume, we have endeavoured to introduce and define our entries on the basis of practical operations, since this is a dictionary of the usage of terms rather than a textbook or a technical encyclopedia. To the purist, some of these words and their applications may appear unusual or even questionable but it is the explanation of current employment rather than grammatical formality which has been our guide. We hope, in addition, that this publication will do something to assist in uniformity of interpretation, especially where the same term has acquired somewhat varied meanings in different contexts.

Our Society is fortunate in being able to call upon so much expert knowledge· in current procedures and practices and gratefully acknowledges the enthusiastic efforts of all the contributors, members of the Dictionary Editorial Committee, many of whom have previously prepared vocabularies as part of their own technical publications. A wide range of other sources has been consulted to provide word lists and we are also indebted to John Halas, Brian Salt and many others, including industry associations, for the basis of many of the definitions which we have included.

In attempting to cover such a wide field of applied technology, and moreover one which is still developing very rapidly, omissions are inevitable and, despite our best endeavours, errors of definition or of usage may have passed uncorrected; readers' comments and additions will be welcomed for inclusion in subsequent editions.

Bernard Happé, *Editor*

BKSTS A-V Dictionary Editorial Committee
Mike A Ray (Chairman)
Stanley W Bowler
Hugh D Ford
John Lewell
Jack Speller
Brian Watkinson
G Ross Watson
David Wilkinson

April 1982

A

a atto-, a prefix denoting a factor of 10^{-18}.

A Ampere, unit of electrical current.

A.2 Antenne 2, the second French state broadcast TV network.

Å *Ångström.*

A-weighting A control of the frequency response of measuring equipment, commonly used for measuring ambient or electrical noise. The network is defined in IEC Recommendation 651.

A/B (1) Monitoring audio signals after or before tape. Now used as a term for comparing audio signals reproduced by different systems. (2) A fading technique between two selected sources.

A & B cutting A method of assembling original film material in two separate rolls, allowing optical effects to be made by double printing. See also *checkerboard cutting*.

A or B types Terms used to identify the emulsion position in 16 mm prints for projection. Type B is run emulsion to lens, type A emulsion to lamp.

A and B windings The two forms of winding used for rolls of film perforated along one edge only; see *Figure A.1*.

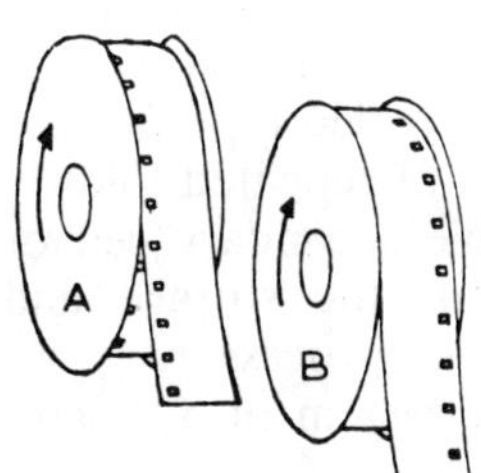

Figure A.1 A or B winding. Film perforated along one edge when unwinding clockwise has the perforations nearer the observer in A winding and away from the observer in B winding

ABC (1) American Broadcasting Company. (2) Australian Broadcasting Corporation. (3) Automatic Beam Control, reduces comet-tail smearing on moving highlights on a television signal by momentarily increasing the camera tube beam current to discharge the highlights.

aberration   Inherent deficiency of an optical system resulting in the formation of an imperfect image. The principal types are *astigmatism, barrel distortion, chromatic aberration, coma, curvature of field, pincushion distortion* and *spherical aberration*.

ABO   Automatic Beam Optimiser = *ABC(3)*

ABS   Association of Broadcasting & Allied Staff (UK).

absorption coefficient   Fraction of incident sound, light or radio frequency energy absorbed by a material at given frequency and conditions.

ABTT   Association of British Theatre Technicians (UK).

ABU   Asian-Pacific Broadcasting Union.

AC   Alternating Current.

Academy   The Academy of Motion Picture Arts and Sciences (USA).

Academy aperture   Aperture of a 35 mm motion picture camera or projector with the dimensions specified by the Academy (= *Academy gate*).

Academy leader   *Leader* on a motion picture theatrical release print containing information and synchronising marks as designed by the Academy.

AC coupled   An electronic circuit capable of passing an AC signal but with a response that does not extend to DC, e.g. transformer or capacitor coupled.

acceptance angle   Horizontal width of vision of a camera.

access   The process of locating and writing or retrieving information in a store.

access time   (1) Time taken to locate and read or write an item of information in a direct access storage device such as a disk drive. (2) Time taken to find a slide in a random access projector, or a desired section on a random access audio or video reproducer.

accordion pleating   Film damage in the form of repeated folds.

accumulator   (1) Electrically, a rechargeable secondary cell or battery of such cells. (2) A register for the storage and manipulation of *operands* and data in a microprocessor.

ACE   Advanced Conversion Equipment, a system of TV *standards converter*.

ace   A 1 kW Fresnel spotlight.

acetate   (1) Cellulose triacetate, a clear transparent plastic used as the base for film and for *cels* in graphics. (2) A metal-cored audio disc on which the modulation is originally cut.

acoustic coupler   A device for data transmission by telephone. It is attached to the phone handset but needs no electrical connection.

action   (1) Command given by a film director for the performance of a scene to begin. (2) The performance of a scene in front of the camera. (3) The film recording the performance, the picture as distinct from the sound.

ACTT Association of Cinematograph, Television and Allied Technicians (UK).

actual sound Sound recorded at the time of filming.

ACTV Advanced Compatible Television, proposed US extended definition TV system compatible with existing *NTSC* equipment.

ACVL Association of Cinema and Video Laboratories (USA).

ADAC Advanced Digital Adaption Converter, a TV *standards converter*.

A-D Analogue-to-Digital.

adaptor A means of interconnecting components of a system having different terminations or couplings.

ADC Analogue-to-Digital Converter.

additive colour Colour mixture by the combination of light of the three primaries, red, green and blue.

address (1) Memory location in computers containing an instruction or data. (2) Selected point in a time-coded tape.

address space The available area defined by the co-ordinates of a computer graphics system.

ADO Ampex Digital Optics. (Trade name).

ADSR Attack, Decay, Sustain, Release—the four phases of *envelope* control in a *synthesiser*.

ADT Automatic Double Tracking. Artificial duplication of signals to simulate two or more musicians, etc.

advance (1) In a motion picture print, the separation between the picture image and the corresponding point on the sound track required for correct synchronisation in projection. (2) The command for an automatic slide projector to advance to the next slide.

aerial An input or output termination for the reception or transmission of radio-frequency oscillations; the American usage is *antenna*.

aerial image An optical image formed in space rather than on a screen. In aerial image photography, titles and other material may be combined with this spatial image.

AES Audio Engineering Society (USA).

AF Audio Frequency.

AFC Automatic Frequency Control, adjusting the receiver to minimise unwanted change.

AFMA American Film Marketing Association (USA).

AFNOR Association Française de Normalisation, the French national standards organisation.

AFPF Association Française de Production de Films (France).

AGB   Audits of Great Britain, for audience research in TV, video and cable (UK).

AGC   Automatic Gain Control, correcting input signal variations to give substantially constant output.

AHD   Audio High Density, an audio disc system. (Trade name)

aiming symbol   A movable *cursor* on a computer graphics display screen, indicating where a *pick* device, such as a light pen or stylus, is pointing.

AIP   Association of Independent Producers (UK).

air brush   A small compressed-air paint spray. The term is also used for a similar effect in computer graphics.

AIRC   Association of Independent Radio Contractors Ltd (UK).

ALC   (1) Automatic Level Control (= *automatic gain control*) (2) Automatic Lamp Changer. (For slide projectors with main and standby lamps).

Al-Ga-As   Aluminium-Gallium Arsenide.

algorithm   A sequence of instructions that performs a given task.

alias   Unwanted signal generated during analogue-to-digital conversion when the sampling rate is too low, typically less than half the highest frequency to be sampled; see *Figure A.2*.

aliasing   The visual effects that occur when the detail of an image exceeds the resolution available on a raster display, for example, a 'staircase' effect in rendering a diagonal line; see *Figure A.3*.

alignment   The process of adjusting a system for optimum performance.

all pass   A circuit without filters restricting its bandwidth.

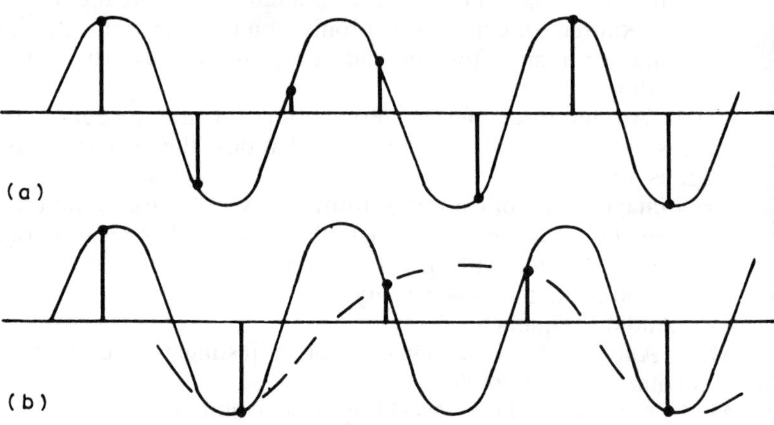

Figure A.2   Alias. In analogue-to-digital conversion (*a*), a spurious frequency is generated if the sampling rate is too low (*b*)

Figure A.3 Aliasing. Because of the raster structure, a diagonal line will be represented as a series of steps

alpha wrap   Tape path in a helical-scan video tape system giving a full 360° contact on the drum; see *Figure A.4*.

alphanumeric   A set of characters that includes the complete alphabet, numerals from 0 to 9, and may include some special characters, such as punctuation marks.

alternate   A facility in slide projection for rapid automatic alternation of two picture images.

alternating current   Electric current reversing its polarity at regular intervals, e.g. the mains supply at 50 Hz in Europe, 60 Hz in USA.

AM   *amplitude modulation.*

ambience   Combination of reverberation and background noise in an environment.

ambient sound   In presentations using multi-channel sound, the sound fed to the loudspeakers in the auditorium.

ambiophony   Creation of artificial reverberation in auditoriums by the use of time delays and multiple loudspeaker systems.

ambisonic   Multiple sound channels providing three-dimensional reproduction.

amplifier   A device for increasing the strength of input signals (either voltage or current) to provide greater power for subsequent use.

amplitude   Level or intensity of a signal, e.g. brightness or loudness.

amplitude modulation   Variation in intensity of a nominally fixed frequency carrier.

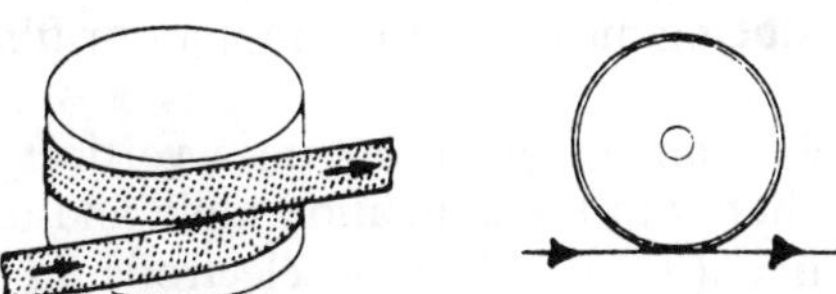

Figure A.4 Alpha wrap. The tape in the helical-scan path makes a complete circuit of the recording head drum

AN *anti-newton.*

anaglyph Stereoscopic presentation in which the right and left eye images are reproduced in different colours, usually red and blue-green, and viewed using a complementary colour filter on each eye.

analogue (analog) A signal continuously proportional to a physical parameter.

analyser, colour See *colour analyser.*

analysis projector Apparatus for the detailed examination of a motion picture film record, frame by frame or at variable speed, often including dimensional measurement of the image.

anamorphic (1) In an optical system, having different vertical and horizontal magnifications. (2) In cinematography, an image having lateral compression produced by an anamorphic lens; see *Figure A.5.*

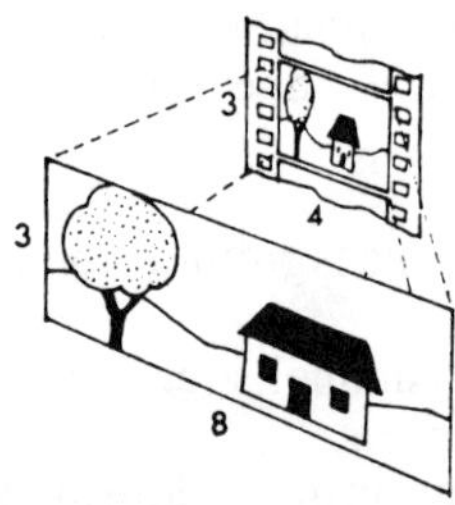

Figure A.5 Anamorphic. The picture image on the film is compressed laterally

AND gate A logic gate which gives an output of 1 when all the inputs are also 1.

anechoic chamber Room enclosed by a non-reflecting surface, may be designed for audio or radio frequencies.

Ångström Unit of wavelength measurement, 10^{-10} metre. (Obsolescent; the preferred term in SI units is the nanometre, 10^{-9} metre)

animate To produce the impression of movement by the rapid presentation of a series of still drawings or slides.

animation Producing movement in graphics or film frames, step by step.

animation stand Camera mounting on a vertical framework or column, together with the animation table and its accessories.

anode A positive (+) terminal or electrode in an electrical circuit.

ANRS Automatic Noise Reduction System. (Trade name).

ANSI American National Standards Institute, Inc., the American national standards authority.

answer print The first print of a completed motion picture production submitted for approval of the laboratory operations.

antenna see *aerial*.

anti-aliasing Processes which remove the effects of addressing individual picture elements on a raster display, making jagged diagonal lines appear smooth.

anti-alias filter A filter before an A-D converter preventing frequencies higher than half the sampling frequency passing to the converter.

anti-bounce A circuit or software provided to remove ill effects, such as mis-counting, which could otherwise be caused by *contact bounce*.

anti-halation A coating or layer applied during manufacture of photographic film to reduce *halation* in the exposed image.

anti-newton Slide mounts with their glass surfaces treated to avoid the formation of *Newton's rings*.

antinode Point at maximum motion in a vibrating system. See also *node*.

Antiope Acquisition Numérique et Télévisualisation d'Images Organisées en Page d'Ecriture, the French national teletext system.

anti-phase Out of *phase* by 180°.

APD Avalanche Photo-diode.

aperture (1) In an optical system, the opening controlling the amount of light transmitted (= *diaphragm*). (2) In motion picture equipment, the opening at which film is exposed or projected; similarly, the exposure opening in a rostrum camera plate. In slide projection, the internal opening of a slide mount.

aperture correction Electronic system for increasing the apparent video picture resolution.

apron A forestage extending beyond the main stage into the auditorium.

APRS Association of Professional Recording Studios (UK).

AQL Acceptable Quality Level.

archive A store with controlled conditions for the long-term preservation of records, including film, audio and video tapes and digital data.

ARD The first West German state broadcast TV network, a consortium of nine regions.

arena An acting area surrounded by an audience.

argument In computing, a value that is passed from a calling program to a function.

array A series of memory locations with a single name, for example a set of related *variables* representing the days of the year could be named DAY (*1*)–DAY (*365*) (a one-dimension array) or named as parts of months DAY (*3, 9*) (a two-dimension array).

artificial light   Light from a source other than natural daylight or sunlight.

artwork   Illustrative copy whether prepared by an artist, camera or other mechanical means. The term is also used in printed circuit layout work.

ARQ   Automatic-Repeat-Request, for automatic error correction.

ASA   American Standards Association, especially its exposure index denoting photographic film sensitivity.

ASC   American Society of Cinematographers, for outstanding directors of photography.

ASCE   Association of Sound and Communication Engineers (USA).

ASCII   American Standard Code for Information Interchange. The standard character code, widely used in the computer industry.

ASFP   Association of Specialised Film Producers (UK).

ASICS   Applications Specific Integrated Circuits, designer chips.

ASIFA   Association Internationale du Film d'Animation, the international association for makers of animated films and video productions.

ASK   Amplitude Shift Keying, a digital modulation method.

ASM   Assistant Stage Manager.

aspect ratio   The proportion of picture width to height, often expressed with the height as unity, e.g. 1.85 : 1, or as 4 × 3 for television.

aspheric   A curved surface of a lens or mirror which does not conform to the shape of a sphere; used for correcting aberrations in optical projection systems.

assemble editing   Editing where material is joined on the end of existing material, without distortion at the edit point.

assembler   A program which converts the low-level mnemonic instructions of *assembly language* to the machine code instructions required to drive a central processor unit.

assembly language   A low-level language using mnemonics to assist program development and understanding. Each mnemonic corresponds to a *machine code* instruction.

AST   Automatic Scan Tracking, in 1″ C-format VTR.

astigmatism   Lens aberration causing points away from the optical axis to be imaged as pairs of lines in different planes.

asynchronous   (1) Not synchronous, e.g. sound not synchronised to the picture being presented. (2) Start-stop data transmission avoiding the need for a synchronous clock at the receiving end.

(3) Two television signals of the same scanning standard but not synchronised to each other, i.e. not *genlocked*.

ATE   Automatic Test Equipment.

ATF   Automatic Track Following, in digital tape or disc reproduction.

ATSC   Advanced Television Systems Committee (USA).

attack time   Time taken for device to respond to the input signal.

attenuator   Device to reduce the strength of an optical, acoustic or electronic signal.

atto-   A prefix denoting a factor of 10^{-18}.

attribute   In computing, a property or characteristic of the specific item(s).

audio   Concerned with sound recording and reproduction; specifically the chain which carries the sound information.

audio cassette   A *compact cassette* used for audio recording.

audio disc   A gramophone record.

audio dub   Recording or re-recording the audio track of a videotape without disturbing the existing video signal.

audio frequency   The frequency range of normal hearing, generally taken as between 15 Hz and 20 kHz.

audio tape   Magnetic recording tape designed to record and reproduce information within the sound spectrum, generally using an analogue signal.

audio-visual   Adjective describing equipment, productions and presentations combining sound and pictures. The restriction of its use to the tape-slide combination is deprecated.

auditorium   The part of a building reserved for the audience.

auto-assemble   Video editing mode in which the videotape recorder spaces the tape back at the end of each recording and runs it up to the start of the next to produce a perfect edit transition.

auto-conforming   Final *on-line* conforming under computer control from a program produced *off-line*.

Autocue   A television prompter. (Trade name).

auto-cycling equipment   (1) In film animation, the automatic photography of one *cel* at specified intervals along the film, the intervening frames being skipped for exposure subsequently. (2) Video or audio tape recorder operating mode in which the tape is automatically rewound and starts again from the beginning or from a pre-set counter position.

auto-focus   (1) In a projector, an electro-optical device to ensure that the image is held in focus. (2) On an animation stand, the equipment which maintains focus as the camera field size is altered. (3) In a camera, a device for measuring the distance of the lens from a given object and automatically setting the lens–film distance accordingly.

automatic end stop A device for detecting a pulse, signal or termination of run and shutting off the equipment.

automatic repeat Keys which automatically continue to repeat their action if held down.

automatic switch-off The interruption of the power supply to the system caused by an end-of-show signal, or by the end of the tape.

auto-stop *automatic end stop.*

auto-threading Film projector, tape recorder or similar device equipped to thread itself instead of having to be laced.

auxiliaries Extra effects controlled during a *multivision* programme. The devices commonly controlled include cine projectors, curtains, fountains, lasers, lights, motors, pyrotechnics and smoke machines.

A-V *audio-visual.*

AVA Audio Visual Association (UK).

AVAMA Audio Visual & Allied Manufacturers' Association (UK).

AVC Automatic Volume Control.

AVD Association of Video Dealers (UK).

A-V format A *compact cassette* track format with mono or stereo on tracks 1 and 2, and the cue-tone or control on track 4, track 3 being unused to improve *crosstalk*. Being superseded by the *IEC* standardisation of tracks 3 and 4, and the space between them, for the control track.

azerty European alternative keyboard layout to *qwerty*, common in France and Germany. It has accents and features a comma rather than a full stop on the numeric keypad.

azimuth The angle between the gap in a magnetic head and the direction of the tape motion; similarly, the angle between the slit in a photographic sound head and the direction of film motion.

B

b *bar.*

B *bel.*

B-format A 1-inch broadcast-quality videotape recorder system.

B-weighting A frequency response weighting network used for measuring noise. The network is defined in *IEC* Recommendation 651 'Precision sound level meters'.

BABT British Approvals Board for Telecommunications, part of *BEAB*.

baby A small spotlight.

baby legs A short adjustable tripod for low-angle shots.

back coating A thin conductive coating applied to the non-magnetic side of recording tapes to improve winding qualities, particularly at high speed.

back focus The distance from the rear element of a lens to the image plane.

backing A coating applied to the base of photographic film to prevent or reduce *halation*.

backing-track (1) Prerecorded audio accompaniment. (2) Sound recording of a combination of instruments and/or vocalists supporting the main performer.

back light Light directed from behind a subject towards the camera, to emphasise subject outline.

back-porch In a television waveform, the brief 'black' period between the end of the horizontal sync pulse and the beginning of the picture information; *colour burst* is transmitted during this period; see *Figure B.1*.

back-projection The projection of a motion picture, still slide or TV image on to the rear of a translucent screen, to be viewed from its front surface.

baffle Panel, usually of a loudspeaker cabinet, in which the individual loudspeaker units are mounted to improve radiation at low frequencies.

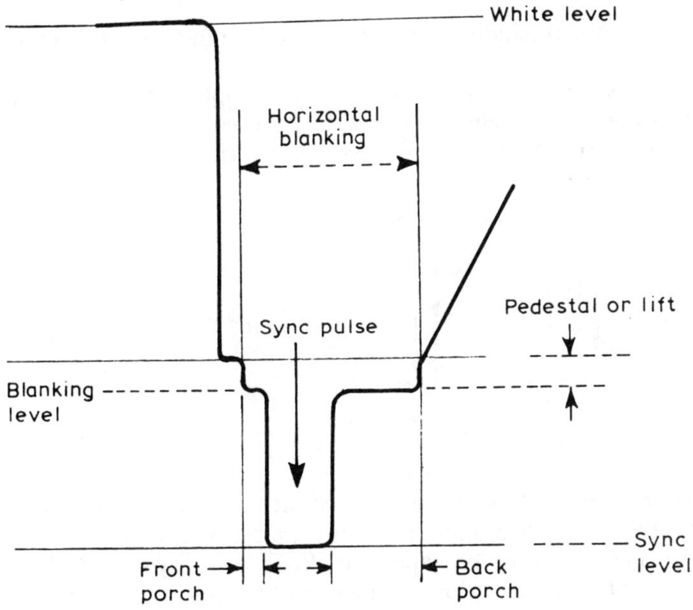

Figure B.1 Back porch

BAFTA   British Academy of Film and Television Arts.

balance, colour   See *colour balance*.

balance stripe   An additional stripe on striped magnetic film to ensure uniform winding.

balanced   Referring to a circuit consisting of a ground connection and two out-of-phase signal lines. Such balanced circuits reject interference common to both signal connections.

banana plug   A single-pole plug, commonly 4 mm diameter.

banding   Also known as *head banding*. A videotape recorder defect which causes horizontal bands of different hue, noise or saturation.

bandpass filter   A filter which transmits a defined frequency range rejecting higher and lower frequencies.

bandstop filter   A filter which rejects a defined range of frequencies passing higher or lower frequencies.

bandwidth   The range of frequencies passed by a device; often specified as the range between the limits at which attenuation of −3 dB below the maximum occurs.

bar   A unit of fluid pressure equal to 100 000 Pascal.

bar chart   A form of graph in which the height of each bar represents the quantity. The correct term is a histogram.

BARB   Broadcasters Audience Research Board Ltd. (UK).

barn door   Hinged flaps restricting the light beam from a *luminaire*.

barney   A soft padded cover used to reduce the noise of a motion picture camera.

barrel distortion   Image distortion in an optical or video system causing a rectangle to appear to have convex sides and compressed corners (cf. *pincushion distortion*); see *Figure B.2*.

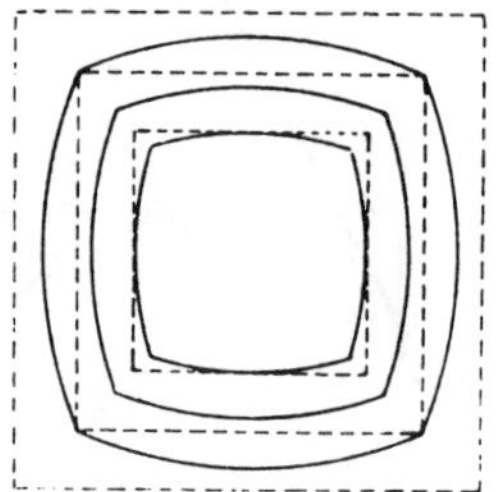

Figure B.2 Barrel distortion

barrier strip   A type of electrical connector comprising a series of screw terminals mounted on, or in, an insulating strip.

base   (1) For film or tape, the flexible support on which a photographic emulsion or magnetic coating is carried. (2) In electronics, the control electrode of a transistor.

basher A small studio *luminaire* placed close to the camera, to which it is sometimes attached.

BASIC Beginners' All-purpose Symbolic Instruction Code. A widely used computer language.

batch number Identification of a quantity manufactured at one time with uniform characteristics, particularly film *raw stock*.

bath A processing solution for photographic film, especially a developer, or its container tank(s).

battery back-up Battery system to prevent loss of data when power to a computer, or other system, fails or is turned off.

baud Unit of data transmission speed representing one change of state per second.

baud rate Speed of data transmission; often but not necessarily equivalent to bits per second.

Baxendall A tone control circuit named after its designer.

bayonet mount A push-and-twist type of lens mounting.

bazooka Adjustable monopod to take a pan-and-tilt head.

BBC British Broadcasting Corportation.

BBFC British Board of Film Classification (UK censor).

BBS Bulletin Board System (USA), an electronic message-passing service.

BCAVM British Catalogue of A-V Material.

BCD Binary Coded Decimal. Four bit or 4-wire system using weighted values of 1, 2, 4, and 8 to express numbers in the range 0-9.

BCU Big Close-Up.

BEAB British Electrotechnical Approvals Board.

BEEAB British Electrical and Electronic Approvals Board.

beaded screen A directionally reflective projection screen, whose surface is composed of minute glass or plastic spherical beads.

BEAMA British Electrical & Allied Manufacturers' Association.

bearding Defect in video reproduction in which the edges of dark areas overflow into adjacent white areas; generally caused by overloading.

beat The cyclic combination in reinforcement and cancellation of two signals at slightly different frequencies, producing a third 'beat' frequency; see *Figure B.3*.

bel Unit used for comparison of magnitudes, W_1 and W_2, on a logarithmic scale; the magnitude difference in bels is $\log_{10}(W_1/W_2)$. The more commonly used form is the decibel, one-tenth of a bel.

BER Bit Error Rate, a measure of digital decoding accuracy.

BETA Broadcasting and Entertainment Trades Alliance (UK).

Betacam Trade name for VTR system of broadcast quality (Sony), using ½″ tape with component recording of *lumi-*

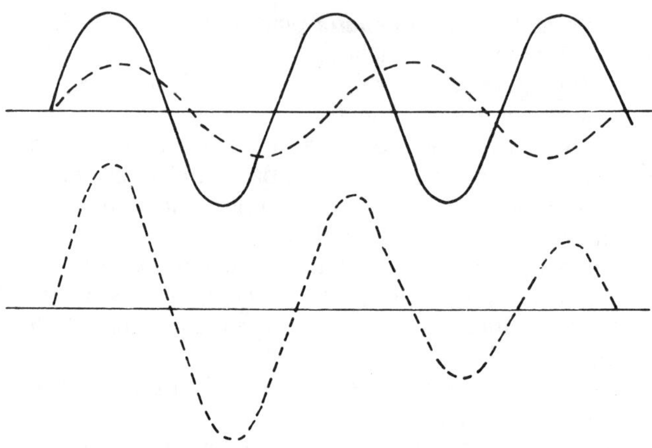

Figure B.3 Beat. Combination of two oscillations of slightly different frequencies produces a third beat frequency

nance and ***chrominance*** on separate tracks; equipment includes TV ***camcorder***.

Beta-format The tape format used in the ***Betamax*** system.

Betamax Video cassette system, using ½-inch tape. (Trade name)

Betamovie ***camcorder*** using standard ***Betamax*** tape format.

BFI British Film Institute.

BFTPA British Film & Television Producers Association Ltd.

bias In audio tape recording, an ultrasonic signal applied to the record head to reduce distortion.

bias trap High frequency filter to remove unwanted high frequency bias from the output or metering circuits of a magnetic recorder.

bidirectional microphone One with normally equal sensitivity to sound arriving from front or rear.

bilateral sound track Variable area photographic sound record whose ***modulation*** is symmetrical about its centre line; may be single, double or multiple.

binary A system of numbering which uses only two states. These are usually expressed as 1 or 0 (one or zero).

binaural sound Sound recorded for headphone listening using two channels which incorporate phase-difference information.

binder The carrier for the metallic oxide particles in a magnetic coating.

binocular vision The ability to use both eyes in viewing a scene.

bipack Two films run in contact through a motion picture or rostrum camera; both may be ***raw stock*** simultaneously exposed or one may be exposed through a processed image on the other.

biphase mark Method used to produce longitudinal *time-code*, adding a correction bit to cancel out any residual DC component.

BISFA Former British Industrial and Scientific Film Association–see *IVCA*.

bistable An electronic circuit with two stable states, usually referred to as 0 and 1; it is the basic element of electronic counting in *binary*, and of electronic memories.

bit A single Binary digIT.

bit plane A computer graphics store for holding a digital representation of a display image as a pattern of bits.

bit slice A form of computer in which the *hardware* has been used in the manner of building blocks to make a central processor unit a chosen number of bits wide, usually in multiples of 2 or 4.

BKSTS British Kinematograph, Sound and Television Society.

black burst A colour TV signal without picture information, with *sync pulses*, *colour burst* and *black level* only; must be recorded onto blank videotape before it can be used for insert editing. (See *Figures B.1, C.6)*

black crush Electronically reduced TV black level, resulting in loss of gradation in shadow areas.

black level The *amplitude* of a TV signal representing the darkest part of the picture, having no exposure.

black level clamp Circuit which holds the black level of a TV picture at a fixed potential or brightness.

black slice see *noise slice*.

blanked vector A vector having no intensity, invoked in a line-drawing display in order to change the current position of the beam without drawing a line. See *vector graphics*.

blanking Period during TV picture formation when the scanning spot returns (a) from right to left after each line (horizontal blanking) and (b) from bottom to top after each field (vertical blanking) during which the picture information is suppressed.

blasting Explosive distortion due to the wind effect of breath on a microphone.

bleed Term used, generally in printing, where illustrations run off the printed page on one, or more, edges.

bleep An audible signal, generally of middle pitch.

bleeper The electrical or mechanical means of producing the bleep signal.

blimp A soundproof housing for a film camera.

blinking Alternately displaying and not displaying a title or object on the screen, usually to draw attention to it.

block A group of characters, words or records treated as a single unit in a computer storage system or in a transmission system.

blocking   Saturation of an electronic circuit inhibiting the passage of signals.

blond   Trade name for a 2000 W variable-beam floodlight.

blooming   Coating the surfaces of a lens in order to reduce reflection losses.

blooping   The technique of applying a special opaque ink, paint or tape. On a photographic sound-track it is a triangular opaque patch to eliminate noise caused by a join.

blow up   To enlarge photographically.

blue backing shot   In special effects, foreground action shot against a uniform blue backing for combination with another scene by *travelling matte* or *chromakey* processes.

BNC   A twist-lock connector widely used for video cables.

board   (1) An assembly of electronic circuits on a single support. (2) A lighting control panel.

boilerplate   American term for *cut and paste.*

boom   Microphone or lighting support set nominally parallel with the floor. Usually held by a vertical stand.

boot (computing)   To get a computer system into an operating condition.

bounce   (1) Vertical unsteadiness of a projected film image. (cf. *weave*). (2) In television, short term periodic variation of *luminance*. (3) Mechanical fault in electrical contacts; see *contact bounce.*

BP   Back projection.

BPI   British Phonographic Industry Ltd.

bps   *bits* per second.

braiding   A collection of fine wires which are plaited together to form a flat cable or to act as a screen in coaxial or multicore cables.

branch   A fork to an alternative sequence of program execution dependant on specified conditions which may be internal or external to the processor.

breakdown   In motion picture laboratory practice, the separation of usable *takes* from the rolls of processed original camera negative.

breakthrough   Unwanted signals appearing in a circuit from other circuits or devices.

breathing   (1) In sound recording, a distracting effect in *compressors* where the background noise level varies with the level of the desired signal. (2) Defect in motion picture camera, printer or projector by which repeated movement of the film in and out of the correct plane causes variations in sharpness.

BREEMA   British Radio and Electronic Equipment Manufacturers Association.

bridge An electrical or electronic instrument for measuring resistance, capacitance or inductance.

bridging Electrically joining two or more points.

brightness The property of a surface emitting or reflecting light; *luminance*. It is measured in candelas per square metre.

brightness control In a video monitor or receiver, the control which increases the *black level*.

broadcast standard In video practice, representing the highest quality of recording and reproduction, capable of meeting the stringent requirements of broadcasting organisations.

broad A *luminaire* for wide general illumination.

brown out A mains voltage reduction, more common on 110 V supplies, liable to lead to equipment malfunction.

BRT Belgische Radio en Televisie, Belgian state broadcast network in Flemish language.

bruch blanking Sequence of switching off the *colour burst* during field blanking in a *PAL* system to improve stability in the colour television receiver.

brush (1) A wiping electrical contact commonly comprising a carbon compound conductor which is pressed on to a smooth conducting surface by a spring. (2) In computer *painting* a marker that draws a line or a pattern on the display surface.

brute A large high-intensity spot light, usually an arc lamp.

BS British Standard.

BSI British Standards Institution. The British national standards authority.

BT British Telecom.

BTL Behind the Lens, filter position in a camera.

BTS Broadcast Television System (a term used in the USA).

bubble memory A non-volatile computer memory where the data is held in the form of magnetic discontinuities or bubbles.

bucket brigade A shift register type of circuit which functions by passing packets of (analogue) information from one element to the next like buckets being passed from hand to hand. Commonly applied to *charge coupled devices*.

buckle switch Switch fitted in a motion picture or rostrum camera to stop it running if the film piles up or is improperly threaded.

buffer (1) A storage area which receives and then releases transient data. (2) Electronic device in a signal path designed to pass or augment signals but to hold back unwanted or damaging conditions.

BUFVC British Universities Film and Video Council.

bug A software error which causes an unintended result. The act of eradicating these faults is termed 'debugging'.

build up (1) In film editing, blank spacing inserted to represent missing sections. (2) Term used for illuminating a row of lights or images, where a new one is added to one end of the row without removing any of those already there.

bulk eraser Apparatus for removing signals from magnetic tape or film in a complete roll, without unwinding.

bulk storage Storage medium, typically *hard disk*, designed for storing large quantities of data.

burn (1) Defective area in a TV camera tube caused by focusing an excessively bright light on it. (2) Damage burnt into the phosphor of a cathode ray tube.

burn-in (1) A technique for producing white or light coloured lettering, logos, etc., by over-exposure on to an exposed image in the camera or by superimposed projection. (2) The addition of time-code numerals visible in the picture to a videotape record. (3) A pre-conditioning technique for components or equipment to improve reliability.

burst Short period of colour subcarrier added to each line sync period. (= *colour burst*).

bus (buss) An electrical link interconnecting items in a system,

butt splice A join in a film or tape in which the two ends are not overlapped. When the ends are fused together, it is termed a *butt weld*.

buzz track (1) A test film with a special photographic sound track used to determine the correct central position of the scanning slit of an optical sound reproducer; see *Figure B.4*. (2) Sound track recorded on open channel with local background noise only, used to fill gaps in commentary or dialogue.

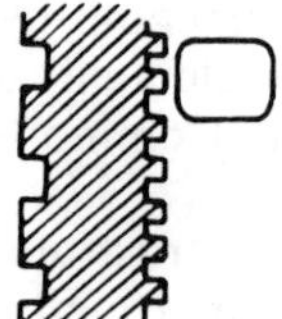

Figure B.4 Buzz track. A test track with markedly different frequencies on its opposite edges to check correct centring in the reproducer

BVA British Videogram Association.

BVU Broadcast Video U-matic. A videotape system, using ¾ inch tape. (Trade name).

B & W (B/W) Black and white, referring to monochrome materials and film or video images.

byte A group of binary *bits*, usually eight.

C

c   centi-, a prefix denoting a factor of one hundredth, 10^{-2}.

C   Coulomb, unit of electrical quantity or electric charge.

C-format   Helical-scan videotape recording system of broadcast quality, using 1-inch tape reel-to-reel.

C-weighting   Frequency response weighting network used for measuring some types of noise. The network is defined in *IEC* Recommendation 651.

C-zero   A *compact cassette* without any magnetic tape.

cable TV   A system for distributing television signals, picture and sound, by means of a cable link.

CAD   Computer Aided Design. The use of the computer to generate or originate graphic material.

CAL   Computer Aided Learning.

calender   To polish a coated surface, for example, magnetic tape.

calibration tape   A pre-recorded magnetic tape containing tones at standardised levels.

call   A computer instruction to execute a *subroutine*.

CAM   Computer Aided Manufacture.

camcorder   Compact hand-held video camera with integral videotape recorder.

cameo lighting   A single foreground subject lit against a substantially uniform dark background.

camera   Device to convert an optical image, formed by a lens system, into a permanent or semi-permanent picture using photographic or electronic techniques.

camera card   Cameraman's cue card giving an individual camera's shots.

camera duplicates   In slide production, the photography of several originals of the same shot to avoid copying by duplication.

camera exposure sheets   Full instructions for every frame for use by the rostrum camera operator in animation photography.

camera original   The film used in the camera to photograph the original scene.

camera ready artwork   Material ready for reproduction by photography.

camera script   Script which includes details of shots, lighting and sound.

candela   Unit of luminous intensity.

cannon connector   An earthed three-pin audio connector.

cans   Headphones.

capacitor microphone   A microphone which operates by virtue of the changing capacitance of a foil diaphram.

capstan   Primary source of constant speed tape drive motion in a magnetic recorder.

Captain   Character And Pattern Telephone Access Information Network system, a Japanese teletext system.

caption   (1) A written or spoken title identifying or explaining the content of a pictorial image. (2) Term for artwork, lettering, photographs or diagrams used for programmes.

caption scanner   Fixed television camera used for captions, often has automated changeover.

CAR   Central Apparatus Room.

card   Removable electronic circuit board with plug-in connector to cardholder or cardframe.

cardioid microphone   A partly directional microphone with a heart-shaped polar response diagram; see *Figure C.1*.

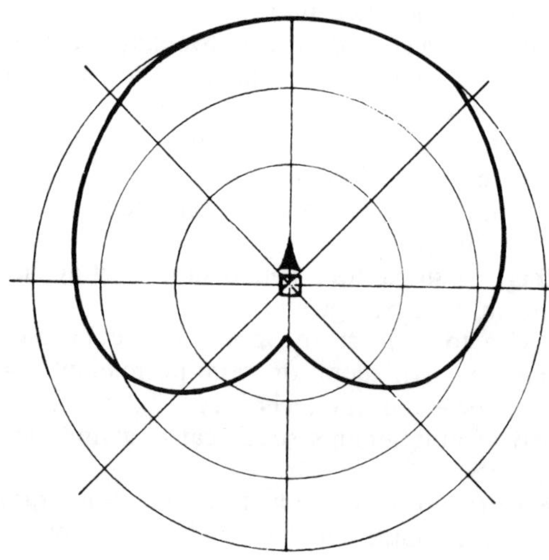

Figure C.1   Cardioid. Polar diagram of the sensitivity of a directional cardioid microphone

card reader   A device for reading data from magnetic or perforated cards.

Carousel   Trade name for a rotary-magazine gravity-feed slide projector, or the magazine itself.

carrel   Working booth for an individual student, usually fitted with A-V equipment such as audio tape recorder, slide projector or video player.

carriage return   The key instructing a computer or visual display unit to execute the contents of its keyboard buffer.

cart machine Equipment at a radio or TV broadcast station for storing and automatically playing a number of audio cartridges or videotape cassettes in a predetermined programme.

cartridge (1) A container for a single spool of film or tape, feeding to a separate spool. (2) A container for a continuous loop of film or tape. (3) The stylus holder and electro-mechanical transducer of a disc pick-up head.

cassette A container enclosing both feed and take-up spools for film or tape.

catadioptric lens An objective system, usually of long focal length, combining transmitting and reflecting surfaces.

cathode The negative (−) terminal or electrode in an electrical circuit.

CATV Community Antenna Television, distributed by cable from a central aerial.

CAV Constant Angular Velocity, e.g. in video disc systems.

CB Citizen Band (Radio transmitter/receivers).

CBA Commonwealth Broadcasting Association.

C-band Radio frequencies 3.7 to 4.2 GHz for satellite use.

CBS Columbia Broadcasting System (USA).

CCD *Charge Coupled Device*.

CCETT Centre Commun d'Etudes de Télédiffusion et Télécommunications.

CCIR Consultative Committee for International Radio; the international radio consultative committee of the *ITU*, the standardising body for radio and television broadcasting.

CCIR Rec. 601 Internationally agreed standard for digital coding of component colour television signal using the 4:2:2 ratio for Y, U and V, with *luminance* sampling at 13.5 NHz and the *chrominance* U and V components at 6.75 MHz; the same frequencies are used for both 525/60 and 625/50 TV systems.

CCITT Consultative Committee for International Telegraphy & Telephones.

CCR Camera Cassette Recorder, alternative term for *camcorder*.

CCT Abbreviation for circuit.

CCTV Closed Circuit Television.

CCU Camera Control Unit. (In the video camera chain)

cd candela, unit of luminous intensity.

CD Compact (audio) Disc. A digital audio recording read by an optical laser system, using a disc of 12 cm diameter or less.

CD-I Compact Disc Interactive, an *interactive* system using compact disc video recordings.

CD-ROM Compact Disc Read Only Memory, a form of disc *videogram*.

CD-V   Compact Disc–Video.

CEA   Cinema Exhibitors Association (UK).

CED   Capacitance Electronic Disc, a grooved capacitance videodisc system.

Ceefax   BBC teletext system (UK).

cel or cell   Transparent plastic sheet used for animation, and for overhead projectors.

CENELEC   European Committee for Electrotechnical Standardisation.

centi-   A prefix denoting a factor of one hundredth, 10^{-2}.

ceramic cartridge   A gramophone pick-up cartridge based on a piezo-electric ceramic.

CET   Council for Educational Technology (UK).

CGI   Computer Graphics Imaging.

chain   A process whereby one computer program automatically invokes another.

Chalnicon   A type of TV camera tube. (Trade name).

changeover   Changing from one motion picture projector to another without interrupting continuity of presentation.

changeover cues   Visual indications, usually dots or circles, near the end of a reel of motion picture film to warn the operator to change from one projector to the other.

channel   (1) A discrete chain carrying a specific signal. (2) An allocated frequency band for transmission of a radio frequency signal.

channel loading   Method of inserting film or tape into a reproduction mechanism or projector by a lateral movement rather than by lacing.

character   A specifiable pattern preparable on a visual display unit, monitor or printer, including letters, numbers and specific symbols.

character generator   Devise to issue the sequence of signals needed to form alpha-numeric characters as captions and titles on a cathode ray tube.

characteristic curve   Graphical representation of the performance of a system, such as the relation between exposure and density in photography or the response of a semi-conductor; see *Figure C.2*.

character set   The sum total of all the characters available in a system. They are not always the same for screen, printer and keyboard.

charge coupled device   A semi-conductor device along which stored information may be moved by charge-transfer. Used as an image sensor in an array of elements in which charges are produced by light focused on its surface.

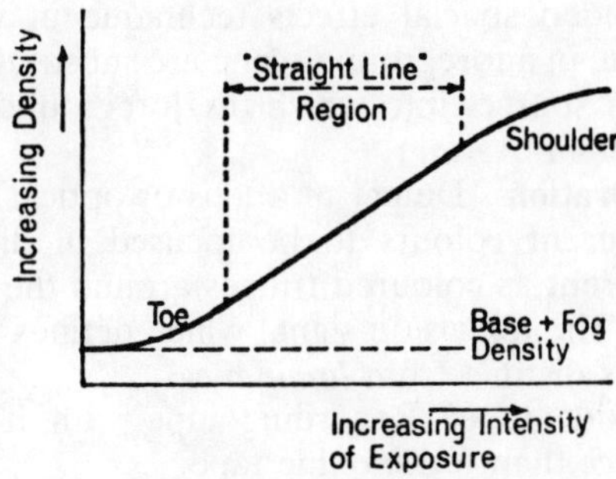

Figure C.2 Characteristic curve. Typical characteristic curve relating the logarithm of the exposure and the resultant density for a photographic material

chase mode   Synchronisation mode between two or more mechanical recorder/reproducers, where slaves follow the master without maximum interlock.

checkerboard cutting   A method of assembling original 16 mm film as alternate scenes in A and B rolls to allow prints to be made without visible splices; see *Figure C.3*.

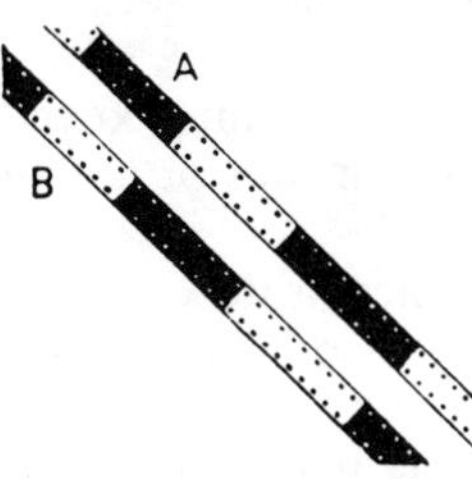

Figure C.3 Checkerboard cutting. Alternate scenes of a reel of 16 mm film are assembled in separate A and B rolls with corresponding lengths of black spacing

checksum   A running binary addition producing a checksum digit at intervals. Used to detect errors in data, typically data received from magnetic media.

chip   (1) An unmounted sub-miniature component. e.g. a chip capacitor. (2) An integrated circuit (slang).

chip camera   Video camera with CCD image sensor(s).

chocolate block   A multi-pole strip connector, often marked for sub-division.

chop   Alternative term for a *hard cut*.

chroma   The component of a TV signal carrying the colour information. Also used to denote degree of colour saturation.

chromakey Video special effects technique in which areas of saturated blue in a foreground scene are substituted by a picture from another source, into which the foreground is 'keyed'. (= *colour separation overlay*)

chromatic aberration Defect of a lens or optical system causing light of different colours to be focused in different planes, usually apparent as coloured fringes around the image.

chrominance The television signal which defines colour hue and saturation, as distinct from *luminance*.

chromium dioxide tape Recording tape with higher *coercivity* and *remanence* than ferric oxide tape.

CID Compact Iodide, Daylight. A light source of 5500 K (Trade name)

CIE International Commission on Illumination.

cinching Uneven winding of magnetic tape on the spool, which may affect the audio/video quality on playback. Also tightening the turns in a loosely-wound roll of film.

cinch marks Longitudinal abrasions on motion picture film or video tape resulting from the movement of one turn against another in a roll.

Cinemascope System of anamorphic wide-screen motion pictures making use of a horizontal compression/expansion factor of 2:1. (Trade name)

cinematography Recording and reproducing motion as a series of images on a strip of photographic film.

cine spool A type of magnetic tape spool with a small keyed centre hole. Commonly made of plastic and 7 inches or less in diameter.

circuit breaker A re-closable electromechanical safety device used to replace the function of a *fuse*; may also be used as a switch.

circuiting The direct transfer of release prints between cinema theatres rather than through an *exchange*.

clapper-board In motion picture photography, a board with a hinged arm used to identify correct synchronisation of picture and sound at the beginning or end of a scene; see *Figure C.4*.

Figure C.4 Clapper board

class A   An amplifier output system which always operates in the linear mode.

class B   An amplifier output system which operates partly in the linear mode.

class C   An amplifier output system which operates outside the linear mode.

class 1, class 2   In the safety classification of mains-powered equipment, Class 1 is earthed and Class 2 double-insulated.

claw   A pulldown device in a motion picture camera or projector which engages and disengages the perforation holes to provide the intermittent motion.

clean feed   Audio monitoring source giving a performer all signals but his own contribution.

clean wind   Full-length fast-forward and rewind of videotape after editing.

click track   A pre-recorded timing track consisting of recorded clicks, used for dubbing music with precise timing.

clip   A short film or video extract from a complete motion picture production.

clipper   A form of *limiter*.

clipping   (1) Waveform distortion in sound or video caused by peak signal amplitude exceeding the available range. (2) In computer graphics, the process of determining which portions of a display element lie outside the specified clip boundary, therefore making them invisible; see *Figure C.5*.

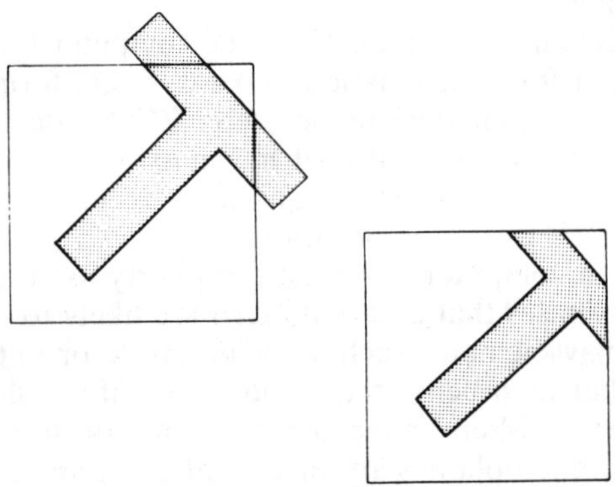

Figure C.5   Clipping. In computer graphics an element is clipped at the boundaries of the display

clock (electronic)   A stable oscillator used for timing, especially in digital systems.

clock pulse   A regular stream of *pulses* often provided for stepping *counters* or *shift registers*.

clock track   (1) A time track recorded on one track of an audio magnetic tape, to facilitate editing and A-V programming. (2) In digital magnetic recording, a timing track necessary for recovering data from the data tracks.

clogging   Build-up of oxide and binder on a tape recording or playback head causing *drop-outs*.

closed loop   A feedback or servo system in which the output result is fed back to be compared with the requirement expressed by the input.

cluster   Assembly of magnetic heads for recording multi-track magnetic tape or film.

CLV   Constant Linear Velocity, e.g. in *videodisc* systems.

C-MAC   Multiplexed Analogue Component coding of a TV signal with the audio signal coded by the 'C' method.

CLV   Constant Linear Velocity, e.g. in *videodisc* systems.

CMOS   Complementary Metal-Oxide-Semiconductor.

C-mount   Standard lens mount for video and 16 mm cameras.

CMRR   Common Mode Rejection Ratio.

cms   Centimetres per second. Strictly cm/s or cm s^{-1}.

CNC   Centre Nationale de la Cinématographie (France).

CNCL   Commission Nationale de Communications et de la Liberté, French regulatory body for broadcasting.

CNR   Carrier-to-Noise Ratio.

cntrl   Abbreviation for *control*.

coaxial   Cable with a central conductor within a cylindrical sheath of screening.

COBOL   COmmon Business Oriented computer Language.

code   To transform information into a different form, usually for efficient storage or transmission, e.g. BCD code.

codec   Coder/decoder for digital transmission.

coercivity   The field strength required to magnetise a material to a given level, usually its saturation.

coherence   In computer graphics, a property used in raster scan which recognises that adjacent *pixels* are likely to belong to the same display element, such as being inside or outside a given shape rather than being on the boundary of the shape.

coincident pair   Microphone arrangement for stereo recording where two microphones are placed adjacent and at 90° to each other. The microphones may be within a single microphone unit.

cold mirror   A dichroic surface reflecting visible light, but transmitting infra-red (heat) radiation in order that it may be safely dissipated.

collector   An electrode of a transistor.

collimated   A beam of light with all its rays parallel.

Colorama Coloured paper available in rolls for studio backings, etc. (Trade name).

colorisation Electronic addition of colour to a videotape transfer of a B&W motion picture for modern TV transmission.

colour analyser Equipment reproducing colour negatives as a positive image by closed circuit television for assessment of the printing levels required.

colour balance The appearance of a colour image considered in terms of the ratio of its three primary colour components.

colour bars Video test waveform.

colour burst A few cycles (8 to 12) of sub-carrier frequency transmitted during the *back porch* period of a colour television signal; see *Figure C.6*.

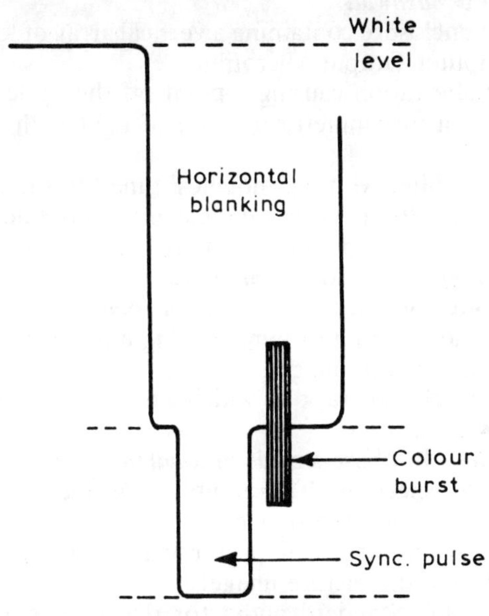

Figure C.6 Colour burst

colour correction Changing the balance or other characteristics in a colour reproduction system to improve the resultant image quality.

colour difference signals The TV signals (B-Y) and (R-Y) which are used to transmit the *chrominance* information.

colour dub A facility provided on some video cassette recorders that provides improved dubbing (copying) performance.

colour filter A transparent material which selectively absorbs specific regions of the visible spectrum.

colour separations A set of monochrome photographic images recording colour information in limited spectral ranges, usually red, green and blue.

colour separation overlay Video special effects technique similar to *chromakey*.

colour temperature A measurement of the colour quality of light. The temperature on the Kelvin scale at which the radiation of a black body matches that of the source in question.

colour under Method of recording a colour TV signal on video-tape, used in most domestic VCR formats; *chroma* is recorded at a lower carrier frequency than the *luminance* signal.

colour wash A facility for the overall colouring of individual screens in a *videowall*.

column An enclosure containing a vertical array of loudspeakers.

COM Computer Output Microfilm.

coma Lens aberration causing a point off the optical axis to be imaged as an unsymmetrical patch of light with a tail like a comet.

comb filter A filter with the 'notches' tuned to the fundamental and harmonic frequencies but leaving intermediate frequencies unaffected. Commonly usd in colour TV decoding to separate *luminance* and *chrominance*.

combined print See preferred term *married print*.

commag A motion-picture film carrying a magnetic sound-track alongside the picture image.

commentary track A track on which a narrative, commentary or voice-over is recorded.

common mode In-phase signals in a *balanced* input.

common mode rejection The ability of *balanced* or *differential* inputs to reject in-phase signals.

comopt A motion-picture film carrying a photographic sound-track alongside the picture image.

compact cassette Standard audio (or data) cassette using tape approx 0.15 inch wide; not to be confused with miniature cassettes used for dictation or data. (Registered design).

compander Combined *compressor* and *expander*.

comparator (1) A circuit element whose output indicates the result of comparing the voltages existing at its two inputs. (2) A computer graphic device which compares the proximity of a *cursor* to the *vector* currently being drawn.

compiler Complex computer program to convert source code (orientated around a high level language) into executable code for later use. The resulting 'object code' is usually, but not always, *machine code*, it can be P-code; see *PASCAL*.

complementary colours Colours resulting from subtracting a given colour from white light, thus the complementary colour to red is 'minus red' which is cyan (blue-green).

component video Transmission and recording colour TV with *luminance* and *chrominance* as separate signals, often with time compression.

composite A photographic *montage* of images, as in *multi-screen*; also, several images on a single slide mount.

composite print See preferred term *married print*.

composite video A television signal with both picture and synchronising information.

compound table A device to provide X, Y and rotary movement used with a *rostrum camera* in animation and in the production of streak and other effects in slides.

compressor Device restricting the dynamic range of programme material to a predetermined narrower range.

computer A electronic device designed to perform a wide range of tasks, utilising some form of flexible programming technique; see *microprocessor*.

computer graphics Visuals originated or animated by computer.

computer language A specific vocabulary and way of using it designed for communication with a computer; see *high level language*.

concatenated code In digital recording, linked blocks of coded information used for error correction.

conceal/reveal Video transition effect in which one scene appears to slide across another: in *conceal* the new scene obscures the previous one, in *reveal* the previous scene moves to display the new scene beneath, (cf. *push*, *scroll*). Also used in *VDU*, where answers are concealed or revealed.

condenser lens A lens, lens system or lens/reflector combination used to concentrate light from a source into a defined beam, as in a projector.

condenser microphone See *capacitor microphone*.

conforming The assembly of picture and sound components in film or videotape to match the editing continuity.

console A control desk for lighting or sound.

constant current A device or circuit capable of maintaining a steady current in spite of changing voltage or varying load resistance.

constant voltage transformer A transformer designed to supply a constant output voltage in spite of a change of input voltage.

contact bounce The pattern of momentary openings and closings which occurs at each operation of most mechanical contacts.

(Even in microswitches, the bounce persists for 1 to 10 milliseconds)

contact microphone Microphone designed to respond to sound vibrations transmitted other than through air.

contact printing Still or motion picture printing in which the processed film to be copied is held in contact with the raw stock at the time of exposure.

continuous animation An *interactive* system of animation as opposed to pre-programmed graphics.

continuous cassette A compact audio cassette containing an endless loop of tape.

continuous printing Motion picture printing in which the processed film to be copied and the raw stock move continuously past the point of exposure. (As opposed to *step printing*)

continuous projection (1) Motion picture projection in which the film moves continuously past the illuminated aperture, the necessary intermittent motion being obtained by optical means. (2) Projection of film as a continuous loop so that it may be presented time after time without a break. A normal intermittent motion projector may be used in conjunction with a loop platter or cassette.

continuous tone masks See *soft-edged masks*.

continuity (1) The correct sequence and matching of all aspects of action and setting between successive scenes in a motion picture or television production. (2) The documents prepared to ensure such results. (3) Link announcements between sequences or between programmes.

continuity still Photograph of a scene taken to record details of setting, costumes, etc., to ensure matching in subsequent shots.

contour correction See *aperture correction*.

contrast The relationship between the light and dark areas in an original or its reproduction.

contrast control In a TV monitor or receiver effectively a gain control for the adjustment of *contrast*.

control Key to be held down to obtain *control characters*.

control characters Non-printing *ASCII* characters used for control functions. They are not displayed on a visual display unit screen.

control track (1) In an audio tape, a track carrying instructions on operations to be performed during running, e.g. mixing, lighting, projection, etc. (2) In videotape, a track used for servo information, synchronisation and scanning rate. (3) In motion picture films with multiple (magnetic) sound tracks, a track controlling the distribution of sound to the various loudspeaker systems.

convergence   Adjustment of electron or optical beams for exact registration of all three colour images.

cookie   An opaque or translucent flat surface cut and mounted so as to provide a desired pattern of shadows on a studio backing.

copier   A device for duplicating images or re-recording sound.

core   A cylinder, usually of moulded plastic, on which a roll of film or tape is wound.

counter   Mechanical, electromechanical or electronic device for counting operations. In motion picture practice, a counter designed to measure the number of frames along the length of the film, expressing the result in frames only, or in feet and frames.

Cox Box   A colour caption synthesiser (Trade name), also used as a generic term for matte generators.

CP/M   Control Programme for Microprocessors, a widely used disk-based *operating system*. (Trade name).

cps   Characters per second (not cycles per second, for which the term hertz should be used).

CPU   Central Processor Unit, in video recording and transmission, and also in a computer.

CR   *carriage return.*

CRA   Camera Ready Artwork, for printing, slides, or overhead projector transparencies. Material ready to be photographed or screened.

crab   Shifting a camera or microphone sideways. Not to be confused with *pan.*

crane   A large camera mounting with a boom arm to lift and support the camera and operator above the scene during shooting.

crash   The failure of a computer to continue to execute the intended instruction sequence; it can be caused by interference or fleeting power failures.

crash editing   Switching a videotape recorder direct from playback mode to record; the new recording may not be in synchronisation with the earlier one.

CRC   Cyclic Redundancy Check, a method of verifying digital data.

crawling title   A line of titles or caption moving horizontally across the screen.

credits   Acknowledgement given in publication, especially the titles at the beginning or end of film or TV productions, listing the cast, technicians and organisations concerned.

creeping sync   In film recording, a progressive error of synchronisation between picture and sound track, and the steps taken for its correction.

creeping titles   See *rolling titles*.

CRI   Colour Reversal Intermediate film, a duplicate colour negative processed by reversal.

crispener   Electronic device for sharpening video picture image edges.

cropping   Cutting off the top and/or bottom of a projected picture to present a wider aspect ratio.

cross compiler   A compiler which operates to produce code for a particular processor while running in a computer utilising a different processor.

cross fade   Transitional dissolve effect between two audio or video signals or between two slide projectors.

cross hatch   Test waveform for making monitor or receiver *convergence* adjustments; see *Figure C.7*.

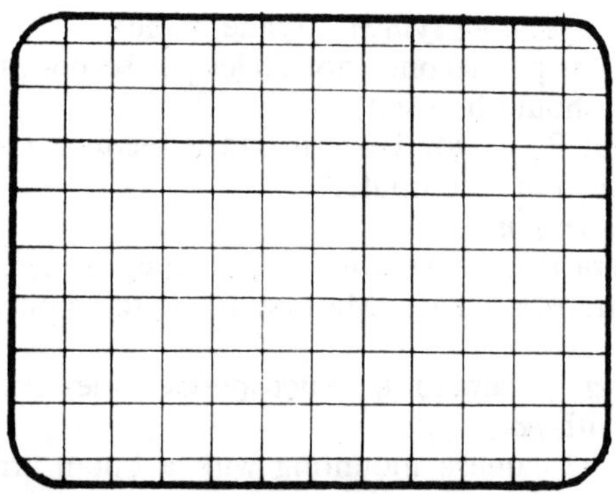

Figure C.7 Cross hatch. Test pattern as displayed on television screen

cross mod   Cross-modulation test, a method for determining the optimum printing requirement for a photographic variable area sound record in motion picture practice.

crossover distortion   A form of distortion in push-pull amplifier circuits which arises as the driving task is passed from one driving device to the other.

crossover network   A filter network dividing an audio channel between two or more loudspeakers.

crosstalk   Unwanted breakthrough between channels of a programme chain.

crowbar   A power-pack protection system, shorting the supply rails if overvoltage occurs:

crow's foot   Notched holder to prevent legs of a tripod slipping.

CRT   Cathode Ray Tube.

crystal   See *quartz crystal*.

crystal sync A method of synchronising an audio magnetic tape recorder to a motion picture camera.

CSI Compact Source Iodide, a particular form of metal halide lamp having a colour temperature approximating to 4300 K. (Trade name)

CSO Colour Separation Overlay, see *chromakey*.

CST Commission Supérieure Technique du Cinéma Français (France).

CTA Cable Television Association (UK).

CTCM Chroma Timer Compressed Multiplex, the basis of the *M II* video system.

CTDM Compressed Time Division Multiple, the basis of the *Betacam* video system.

CTR (1) Current Transfer Ratio. (2) Cassette Tape Recorder (audio).

CTRL See *control*.

CU Close-up shot, typically showing only the head of the subject.

cue A command or signal for a previously specified event to take place, for example, action or speech to commence or for a device to carry out the next item of its programme.

cue/review A control on an audio tape recorder, to keep the head in contact during fast winding.

cue dots Visible signals to indicate the end of a programme section. In film, usually white circles in the top right-hand corner of the frame for projector run-up and change-over; in television, usually a white square in the top right or left corners.

cue tone An audio frequency of specified duration recorded on the cue track so as to provide the indication of the prompt or cue.

cue track (1) A control track used for tape-slide synchronism or control purposes on audio tape recorders and cartridge machines. (2) A secondary audio track on a video tape, available for electronic labelling or for a secondary programme channel.

cukuloris See *cookie*.

current loop In data transmission, a serial transmission system where current flowing represents a mark and no current represents a space.

cursor The character indicating the currently active position on a visual display unit screen.

curvature of field Lens aberration in which the sharpest image is formed in a curved surface rather than a plane.

curved-field lens A projection lens designed to focus a curved surface rather than a plane. Used to project cardboard mounted slides which have 'popped'.

cut (1) Command given by a film director to stop the performance of a scene and its photography. (2) In editing, an instantaneous change from one scene to another. (3) In slide projection, a rapid change from one image to the next produced by lamp control. In order of increasing speed they may be termed, Cut, Fast Cut and Hard Cut; the very fast *snap* is produced by shutters.

cut and paste (1) In computer graphics, moving an item to a new location on a display. (2) A word-processing term for the facility for combining blocks of standard text with variable inserts to create personalised documents, such as mailshots.

cutaway Non-critical shot used to break or link principal action in scenes.

cut-out (1) Electrical protective device to break a circuit on overload or overheat. (2) In animation, small pieces of artwork cut to shape for movement frame by frame during photography. (3) A flat composition cut to outline, used in studio set construction.

CUTS Computer Users Tape System. A 'standard' used for recording data on to compact cassettes.

cutting frames Extra frames at start and finish of an animation scene to give latitude in editing.

CVC Compact Video Cartridge, using ¼ inch tape. (Trade name).

CV-One A video recorder (Hitachi) using ¼ inch tape with component format for separate *luminance* and *chrominance* tracks.

CVR Cartridge Video Recorder, using ½ inch tape.

cyan The complementary colour to red.

cycles Set of artwork for an action which repeats itself after a certain number of frames.

cycle-time see *dissolve cycle time*.

cyclorama A smooth curtain or back cloth suspended around the periphery of a studio or stage.

D

d deci-, a prefix denoting a factor of one tenth, 10^{-1}.

D-1 Digital VTR system using component signal recording to *CCIR Rec. 601* on ¾ inch tape cassette with four digital audio tracks.

D-2 Proposed digital VTR system (Ampex) using composite signal recording sampled at four times the sub-carrier frequency on ¾ inch tape cassette with four digital audio tracks. (Not to be confused with **D2-MAC**)

da deca-, a prefix denoting a factor of ten times, × 10

D&AD Designers and Art Directors Association (UK).

D-A Digital-to-Analogue.

DAC Digital-to-Analogue Converter. A device which converts a coded digital signal into its analogue equivalent.

DAD Digital Audio Disc.

dailies In motion picture practice, the first prints made from newly processed picture or sound negative to check content and quality. (Also called *rush prints* or *rushes*).

daisy wheel A medium-speed high quality printer where the type is carried on the spokes of a wheel. The typeface can be changed rapidly.

damping Decreasing the amplitude (usually exponentially) of an oscillating system with respect to time.

Critical damping occurs when a steady state is reached after only one cycle of oscillation. *Over-damping* takes less than one cycle. *Under-damping* takes more than one cycle. See *Figure D.1*.

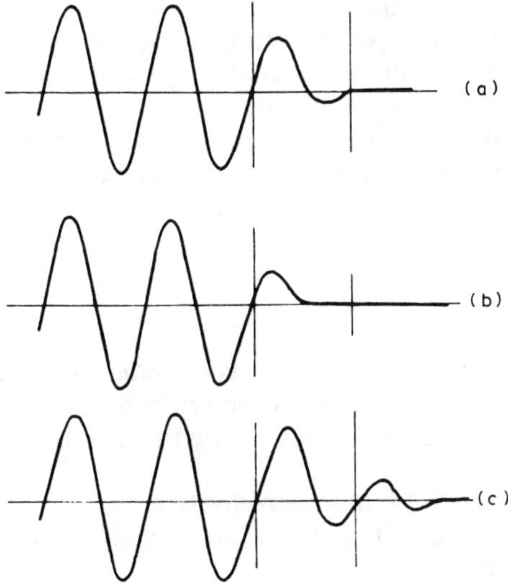

Figure D.1 Damping. Critical damping (*a*) is complete in one cycle of oscillation, while over-damping (*b*) and under-damping (*c*) require less and more than one cycle respectively

damping factor   The nominal load impedance of an amplifier divided by its actual output impedance. A measure of the damping effect upon the loudspeaker.

darlington   Two transistors in a cascade configuration. May be in a single package.

DASH   Digital Audio Stationary Head, audio-tape recording system.

dashpot   A mechanical damping device which may consist of a body moving in a viscous medium.

DAT   Digital Audio Tape recording, *R-DAT* with rotating head, *S-DAT* with stationary head.

data   Alphanumeric or other information fed to or from a computer or digital device.

database   A system for organising a store of information or data in a specific area which can be accessed by a computer.

data projector   A high definition video projector, monochrome for data only or **RGB** for data and video.

data tablet   The flat working surface used in computer graphics for the input of information, with a *stylus* for drawing or a *puck* for digitising.

DATV   Digitally Assisted Television, a European proposal for reducing *HDTV* bandwidth for transmission by a single *DBS* channel.

daylight   Illumination by natural sun and sky or by a source of equivalent colour temperature, generally between 5000 and 6000 K. Colour film for daylight exposure is normally balanced to 5500 K, 'average' sunlight.

daylight projection   Screening of images in high ambient light, requires a high-gain screen with a narrow viewing angle, but back projection may be preferred.

daylight screen   A projection surface of high reflectivity (and hence narrow angle of view) for use in high ambient light conditions.

dB   *decibel*.

dBA   Decibels measured with *A-weighting*. Commonly used for measurement of ambient or electrical noise.

dBm   Power level in decibels related to 1 milliwatt in a 600 ohm circuit.

DBMS   Database Management System.

DBO   Dead Black-Out. Total and usually sudden darkness on stage or studio.

DBS   Direct Broadcast Satellite.

dBV   Voltage ratio in decibels relative to 1 volt.

dBW   Power ratio in decibels referred to 1 watt into an 8 ohm load; used in audio amplifier specifications.

dBx   A noise reduction system that encodes a signal by compressing to a 2:1 ratio. After storage or transmission, the signal is expanded in a 1:2 ratio to restore its original dynamic range. (Trade name).

DC Direct Current.

DC coupled Capable of transmitting signals down to and including DC.

DDC Digital Dynamic Control, used in audio processing.

DDR Direct Domestic Reception (of television transmission by satellite).

debugging Locating errors in an electronic system or computer program.

deca- A prefix denoting a factor of ten times, $\times\ 10$.

deci- A prefix denoting a factor of 10^{-1}.

decibel (dB) Logarithmic measure of relative intensity of power or voltage. Equal to one tenth of a *bel*.

deck (1) The turntable and drive for an audio disc. (2) For magnetic tape, the tape transport mechanism with the heads and associated electronics, generally without the power amplifiers.

decoding The operation of restoring coded data to its original form.

decoder A device which decodes signals. (cf. *encoder*).

decoupling The use of components (usually capacitors) to reduce unwanted coupling of signals from one part to another of an electronic circuit. In particular power supply feeds require to be decoupled.

dedicated Equipment designed to do one task, e.g. a purpose-built word processor, compared with a general purpose computer running a word processor program.

de-emphasis Restoration of the frequency response after transmission or recording which involved *pre-emphasis*.

default value In computer usage, the value which will be assumed in the absence of a specific instruction; for example, in word-processing, the number of lines per page could be assumed as 66 in default of entering a figure.

deferred In computing, an instruction not to be executed immediately, but only when the program is run.

degausser Device for reducing residual magnetism, for example in tape heads.

delay line A device for storing data by transmitting it down a line which is slow in the mode of propagation used. Sometimes the storage is only for the purpose of providing a delay.

delimiter A character used in computer practice to separate or isolate segments of a command.

delta modulation A form of modulation where the latest change of data is used rather than the latest data.

demodulator A device for extracting the wanted signal from a carrier wave.

density Measure of light attenuation, for example of a photographic image, being the logarithm to the base 10 of the opacity, the reciprocal of transmission.

densitometer An instrument for measuring the optical density of media.

depth cueing Computer graphic techniques that are used to give an appearance of depth in a display by varying the intensity levels of the displayed elements.

depth of field The range of object distances from a camera within which objects will be reproduced as acceptably sharp images.

depth of focus In an optical system, the range of distances from the lens within which the image is acceptably sharp.

deuce A 2 kW Fresnel spotlight

developing The process by which the latent image in an exposed photographic film is made permanently visible.

deviation The range of frequencies over which the carrier frequency is swept or modulated in a frequency modulated system.

DGA Directors Guild of America (USA film & video)

dialect A variation of a 'standard' programming language.

diaphragm (1) The opening in an optical system controlling the amount of light transmitted; the transmission of an *iris diaphragm* is adjustable. (2) A membrane used in certain microphones and loudspeakers.

diapositive, (dia) French and German word for a transparency.

diascope A still transparency projector. (Obsolete term).

diazo An imaging process based on diazonium compounds used in overhead projector transparancies, slides, prints and the preparation of photo-resist materials.

DICE Digital International Conversion Equipment, an early system for broadcast TV *standards conversion*.

dichroic Having selective reflection and transmission for radiation of certain wavelengths. A dichroic filter or mirror will transmit a given spectral range and reflect others, as in the colour beam-splitting system of a TV camera or the separation of visible light and infra-red (heat) in a projection system.

dichroic head A colour light source based on adjustable dichroic filters, used with enlargers and rostrum cameras.

dielectric An electrical insulator.

difference frequency The sum or difference between two fundamental frequencies. Difference frequencies are produced when the fundamental frequencies encounter non-linearity in a system.

differential amplifier   Device to amplify the difference between two input signals.

differential input   An electrical input system having two inputs of equal sensitivity combined in anti-phase, giving an output proportional to the difference between the inputs.

differentiator   An electronic circuit which derives the rate of change of its input signal.

diffraction   (1) A deviation in the direction of a wave at the edge of an obstacle in its path. (2) Interference patterns produced as a result of radiation (sound, light or radio) passing around an opaque obstruction.

diffuser   A translucent filter placed in front of a light source to soften the resultant shadows, or in front of a camera lens to reduce the sharpness of the image.

diffusion   Scattering of light.

digit   A single number to any base.

digital   Expressed in numerical terms, often in coded form, as opposed to *analogue.*

digital delay   A signal delay system which works by storing numbers representing the signal.

digital pitch changer   A device using digital techniques to alter the pitch (frequency) of the original sound. Used with only a small pitch change to reduce the onset of *threshold howl.*

digitizer   A device to aid input of digital information, e.g. dimensions from a drawing.

DIL   Dual-In-Line, packaging of an integral circuit with the mounting and contact pins in two parallel rows for flat mounting.

dimmer   Electrical or electronic device for reducing lamp intensity.

DIN   Deutsches Institut für Normung, German national standards organisation.

DIN plug   A range of milti-pin connectors standardised by DIN. The circular DIN connectors and the loudspeaker connector are frequently found on A-V equipment.

diode   Electronic component which allows transmission of electricity in one direction only.

dioptre   (1) A unit of magnifying power of a lens, being the reciprocal of its focal length in metres. (2) A term for a single supplementary lens used in front of the main camera lens for close-up photography.

dip   Metal trap in floor for electrical sockets.

dipole   Aerial (antenna) having two elements of equal length, fed from the centre.

direct cut   An audio disc cut 'live' without the intermediate use of magnetic recording.

disable   A control input. If it is 'true', the circuit is disabled, or not permitted to function.

disc   (1) An audio recording on a flat circular rotating medium. It may be an analogue recording by means of a mechanically modulated groove, or a digital disc using optical or other means of recording. (2) See *videodisc.*

dish   A concave aerial for microwave transmission or reception. Also a reflector used with one form of directional microphone.

disk   In computer practice, a *hard disk* or a *floppy disk* with a magnetic coating used for bulk storage of digital data.

disk crash   Computer failure resulting from a hard disk head actually touching the disk surface; head and disk can suffer physical damage.

diskette   A *floppy disk.*

disk file   Computer data stored on a magnetic disk under a file name.

dispersion   (1) Separation of light into its constituent wavelengths, as in white light passing through a prism. (2) The process of mixing the magnetic particles and other constituents of the coating during the manufacture of magnetic tape.

display   The characters and graphic shapes on the cathode ray tube screen of the visual display unit.

dissolve   A visual transition between two pictures in which the whole area of the first gradually disappears as it is replaced by the second.

dissolve cycle time   In slide projection, the dissolve time plus the time taken for the next slide to be positioned ready for projection.

dissolve time   The period which the transitional change appears to take.

distortion   (1) In electronics, the generation of unwanted signals such as harmonics or intermodulation products. (2) A change, generally deleterious, in the quality of a reproduction as compared to the original.

dither   (1) The intentional addition of low level noise in a digital recording system, to reduce quantising effects. (2) In a mechanical system, an intentional introduction of a small movement to overcome sticking. (3) In computer graphics, an *anti-aliasing* technique used in raster displays in which colour or intensity variations are introduced.

DMA   Direct Memory Access. A method of fast direct input or output of data to or from a computer memory, without the use of the central processing unit.

D-MAC, D2-MAC   Multiplexed Analogue Component coding of a colour television signal, with the audio signal coded to the *D-1* or *D-2* standard.

DMM   Digital Multi-Meter.

DOC   Drop-Out Compensator, providing a signal to replace missing information caused by momentary loss on a video tape.

dock   Area used for temporarily storing scenery.

Dolby level   A reference modulation level used to adjust the Dolby noise reduction system; a fluxivity of 185 nWb/m on magnetic tape or 50% modulation in optical recordings.

Dolby system   A noise reduction system for magnetic and photographic sound recordings or FM radio transmission. (Trade name). *Dolby A* is the professional system which splits the audio spectrum into four frequency bands. *Dolby B* is a domestic system used for cassette tape and FM radio. *Dolby C* is a more recent domestic system for cassettes, and offers more noise reduction than Dolby B.

dolly   Movable platform on to which a camera may be mounted so that action before it may be followed, hence dollying; see also *tracking*.

dope sheet   Breakdown of instructions for shooting scenes; see also *story board*.

Doppler effect   The perceived change in wavelength where there is relative motion between a wave-energy source and an observer. Sound, for example, suffers a change in pitch, light a change in colour, and RF radiation a change of frequency.

DOS   Disk Operating System. A computer program for hard or floppy disk access management.

dot matrix   Alpha-numeric characters formed within a standard pattern of dots.

dot matrix printer   A printer which marks the paper by pins impinging through an inked ribbon.

double density   A system which records (usually on floppy disks) with a higher density to get more data on a given size of medium.

double duplex sound track   Photographic sound track having two *duplex* variable area records side-by-side in phase, within the standard track width.

double exposure   Exposure of the same piece of film twice to receive two different images, which may be superimposed or may occupy areas reserved by *mattes*.

double frame   A system of motion picture photography in which the image area is twice that generally taken as standard.

double frame animation   Each drawing or diagram in a sequence is photographed twice, usually to save labour.

double-headed   Film projection or similar equipment capable of accepting separate picture and sound films, and running them in synchronism.

double-sided sound track   Photographic sound record with a different sound track on each edge, recorded in opposite directions. (= *double-edge sound track*)

double system   In cinematography, the system in which picture and sound are recorded on separate films or on film and tape.

double tracking   Second identical performance of previously recorded video line.

down rate   In slide projection, the time taken from the initiation of a fade, to the disappearance of the light of the down-going projector.

downstage   Performing area nearest the camera or audience.

downstream keyer   A caption superimposer on vision mixer output, which is not affected by the previous mixer controls.

downtime   The time for which a system or unit is inoperative due to maintenance or repair.

dowser   A shutter used to cut off the light beam in a projector or film printer.

DP   Data Processing.

DPCM   Differential Pulse Code Modulation.

DPE   Digital Production Effects. (Trade Name)

DPM   Digital Panel Meter.

dragging   Moving a computer graphics display item along a path determined by a graphic input device; see *Figure D.2*.

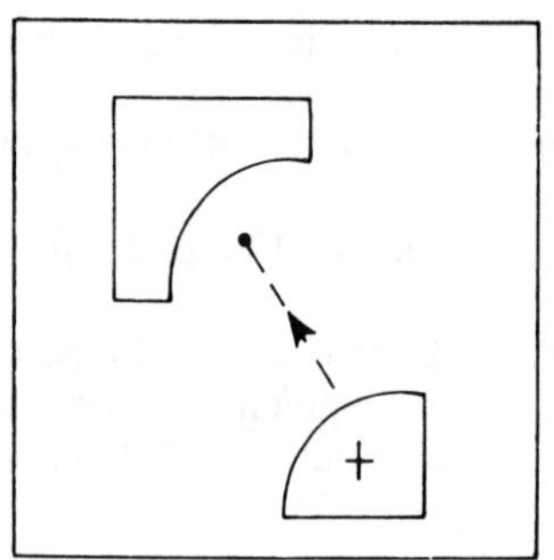
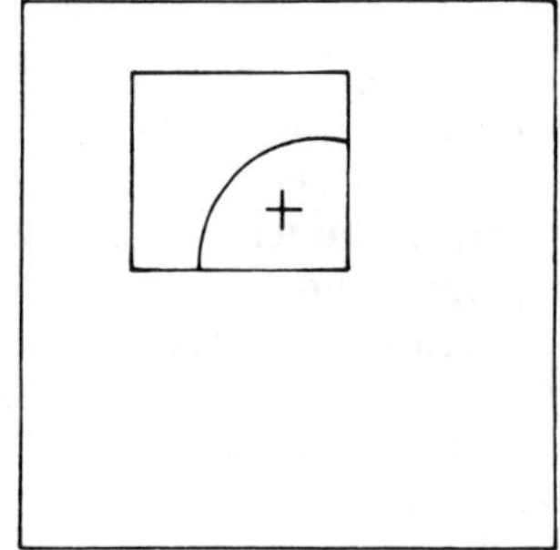

Figure D.2   Dragging. Moving a computer graphics element along a specified path

DRAM   Dynamic Random Access Memory.

drapes   Curtains.

draw   The generation of a computer graphics *vector* by specifying a new endpoint.

DRAW   Direct Read After Write, in a data system such as a recordable video disc.

drift   The slow variation of a parameter, e.g. oscillator frequency or tape speed.

drive   In computer practice, usually refers to a disk drive but strictly means the transport mechanism for any type of large-scale storage medium.

drop-in   In audio recording, the inserting of a replacement section into an already recorded piece by switching the required channel to record at the correct position along the tape. Similar to *insert* in video.

dropout   A short loss of signal in a magnetic recording system due to loss of head-to-tape contact or faults in the magnetic medium.

dropout compensator   Device which reduces the subjective effect of dropouts on a reproduced video tape, by repeating a previously stored line.

drop shadow   Dark edge or edges to lettering and other objects giving an appearance of relief; the effect may be produced photographically or electronically as well as in the original artwork.

drum   (1) A rotatable cylinder round which film passes to control its uniformity of movement in a photographic sound reproducer. (2) Mechanical assembly containing the video record/replay heads in a helical-scan videotape recorder.

dry hire   Hire of equipment without operating personnel.

dry loop   A circuit containing a normally open relay or switch contact which is isolated from other parts of the system and from ground.

D/S   (1) Double-sided. (2) *Downstage*.

DSM   Deputy Stage Manager.

DSP   Digital Signal Processing. Processing a signal (typically audio) in real time by means of a microprocessor.

DTC   Dual Time-Code, videotape recording system to provide additional information in editing.

DTF   Dynamic Track Following, used in high-band videotape replay to obtain perfect tracking in slow-motion or still frame.

DTI   Department of Trade and Industry (UK).

DTL   Diode-Transistor Logic.

DTP   Desk-Top Publishing, using word-processor equipment.

dual cone   Loudspeaker with two cones driven from one voice coil.

dualateral sound track   Photographic sound track with two in-phase unilateral variable area records with modulations in the same direction.

dubbing   (1) Combining two or more sound records into a single composite recording. (2) Transferring a sound recording from one medium to another, e.g. from magnetic tape to photographic film. (3) Making a copy, e.g. a show copy cassette from a reel-to-reel tape. (4) In motion picture production, the process of recording new dialogue to be substituted for the original version.

dubbing switch A switch on video cassette recorders used to improve the quality of tapes being copied.

dumping Saving data from electronic memory on to a permanent medium, for example to floppy disk or cassette tape.

dupes Abbreviation for *duplicates*.

duplex A mode of data transmission where each station can send and receive simultaneously.

duplex sound track Photographic sound track with two in-phase unilateral variable area records having modulations in opposite directions.

duplicate In general terms, an exact copy as distinct from an original record. (1) For slides and for audio and video tapes, a reproduction matching the original as closely as possible. (2) In motion pictures, duplicate refers to a copy negative matching the original exposed in the camera.

duplication In photographic practice, the preparation of a film or slide duplicate from original material. In audio and videotape usage, the preparation of copies in general.

DVE Digital Video Effects.

DVM Digital Voltmeter.

DVR Digital Video Recording.

D-weighting Frequency weighting network sometimes used for the measurement of aircraft noise.

dyne Unit of force. That force which will accelerate a mass of 1 g at 1 cm s^{-2}.

dynamic Data memory where data is lost unless constantly refreshed, by recirculating the data or in some devices just by addressing it.

dynamic microphone A moving-coil type of microphone.

dynamic range The usable range between the strongest signal capable of being handled by a device and the inherent noise.

E

EAROM Electrically Alterable Read-Only Memory.

EAVA European Audio-Visual Association.

earth A conductor connected to ground. (The American usage is *ground*).

EBU European Broadcasting Union.

ECL Emitter Coupled Logic.

echo The return of 'characters from keyboard to monitor screen: local echo direct from own origination, remote echo from a distant system.

echo unit In audio recording, equipment for generating time-delayed signals to be added to the original sound.

edge numbers Groups of sequential numbers, sometimes including letters, along the edge of motion picture film at regular intervals, usually one foot apart.

edge triggered A *monostable* or *bistable* which is triggered by a specified direction of change in the input state, rather than by the actual input 1 or 0, e.g. triggered by the rising edge of the waveform.

edgewave Film distortion in which the edge is stretched longer than the centre.

editing The process of selecting from a large amount of recorded material that relevant to the purpose and assembling a programme by transposing or combining separate sequences.

editing block An aid to joining magnetic tape.

EDL Edit Decision List, giving the source and time-code of videotape edits.

EDP Electronic Data Processing.

EDTV Extended Definition Television, proposed system (Hitachi) for higher resolution by interleaving but compatible with existing 525- or 625-line bandwidth.

EECO Videotape recorder editing equipment. (Trade name).

EEPROM Electrically Erasable Programmable Read-Only Memory.

EETPU Electrical, Electronic, Telecommunications and Plumbing Union (UK).

effects track A sound track containing sound effects only.

EFP *electronic field production.*

egg boxing The division of a multi-image screen into compartments so as to provide a sharp edge between each projected section (= *egg crating*).

EIA Electronic Industries Association of USA. From this, the US television standard which has 525 lines with 60 fields (30 frames) per second.

EIAJ Electrical Industries Association of Japan.

Eidophor A powerful TV projection system using light from a separate source passing through a film of oil which is distorted by the scanning electron beam. (Trade name).

eigentones Preferential frequencies at which acoustic resonances occur, related to parallel walls in a room.

ELCB Earth Leakage Circuit Breakers, a safety device which disconnects line power if current flows to earth.

electret microphone A microphone similar to a capacitor microphone but with a permanent charge on the foil so that it does not need a polarising voltage supply.

electromotive force The electrical potential measured in volts.

electron gun   The device at the rear of a cathode ray tube which provides a defined beam of electrons.

electronic balanced input   A system where the balanced circuit is obtained using active components rather than centre-tapped transformers.

electronic editing   The selection and composition of video information, making a new recording by a transfer from a variety of original material.

electronic field production   Location recording, usually for drama, using lightweight television cameras and recorders, to full broadcast standard.

electronic news gathering   Recording of newsworthy subjects using lightweight, battery-powered television cameras and recorders. Picture quality may have to be slightly downgraded from studio quality to achieve weight and power reductions.

electronic publishing   Images reproduced through television, video devices, discs, tapes, cassettes, closed-circuit broadcasts, cable broadcasts and other electronic means.

elevator   An arrangement of fixed and moveable film rollers acting as a variable capacity reservoir for a film processing machine.

EMF   *electromotive force.*

emitter   An electrode of a transistor.

emitter follower   A transistor connected to provide a low output impedance with approximately unity gain.

emulsion   In photographic materials, the layer containing the light-sensitive agents in a medium such as gelatine.

enable   A control input. If it is true the circuit is enabled or permitted to function.

encoding   (1) Putting information into a prescribed coded form, suitable for transmission. (2) In slide projection, adding control *cue tones* or signals to an already prepared audio tape.

encoder   Device for encoding information, for example a multiplexing 8-wire to 3-wire binary encoder. A shaft encoder converts angular position or revolutions into digital form.

encryption   Scrambling a TV transmission so that it can only be reproduced at stations equipped with a corresponding decoder, as applied in *payTV*.

endless cassette   An audio cassette containing an endless loop of tape, which can run only one way and cannot be rewound.

ENG   *electronic news gathering.*

engineering cue   An additional cue provided with certain tape-slide systems to advance the projector magazines to the first programme slide.

engineering pulse   A recorded command, cue tone or signal, which checks the operation of a system, sets the parameters of operation or advances the equipment to the start position.

enhancer   In television, a device to increase the apparent resolution, either horizontal or vertical.

envelope   (1) The enclosure containing the components of a lamp, cathode ray tube or similar vacuum device. (2) The boundary of the excursion of amplitude of a signal, such as a variable-area sound record or an RF signal; see *Figure E.1*.

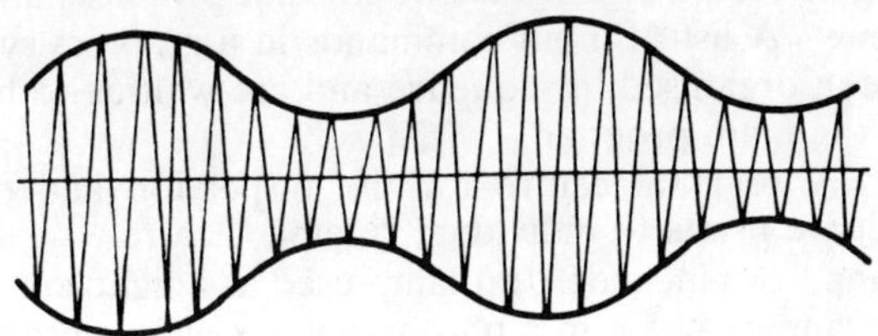

Figure E.1   Envelope. The boundaries of the amplitudes
of a signal

ePAL   Enhanced *PAL*, a modification of the PAL colour television system to improve resolution and reduce colour interference on checkered patterns.

epidiascope   An optical projector which combines the function of a transparency projector (diascope) with that of an *episcope*.

episcope   An optical projector producing an image of opaque objects, such as photographs or pages of a book, by reflection from their surfaces.

EPP   Electronic Post-Production. See *post production.*

EPR   Electronic Pin Registration, ensuring consistent frame-to-frame steadiness in *telecine* transfer.

EPROM   Erasable Programmable Read-Only Memory.

equaliser   Device for altering the frequency response characteristic of a signal.

equalisation   The use of frequency-selective attenuators and/or amplifier circuits to change the balance in amplitude between signals in the upper and lower frequencies.

erase   To remove a previous recording from magnetic tape or film.

eraser, bulk   Device for removing signals from magnetic tape or film when wound in a complete roll, as in a cassette.

eraser, UV   A device to erase *EPROM* ready for re-programming.

error message   Response from computer system to corrupted or incorrect data, etc.

ERT   Elliniki Radiophonia Tikleorassi, Greek state broadcast network.

ETA   Educational Television Association (UK).

E-to-E   Electronic-to-electronic.

ETSA   European Television Services Association.

ETV   Educational Television.

Eureka EU95   European proposal for broadcast *HDTV* using 1250 lies, 50 fields per second, compatible with existing *PAL* receivers.

event   In computer graphics, an action by the operator, such as pressing or releasing a *stylus* switch, that produces an input.

event queue   A list of input commands in a graphics system that have been organised in sequence and are waiting to be processed by the main program.

exchange   A regional centre for the inspection and despatch of film release prints to individual cinema theatres.

exciter lamp   An incandescent lamp used as a light source for the sound scanning beam in a photographic sound reproducer.

EXCLUSIVE-OR gate   A logic gate where the output is true when either, but not both, inputs are true. Hence it is sometimes called a non-equivalent gate.

excursion   The limits of variation of a signal or motion.

execute   In a computer, to obey and act on the instructions contained in the program.

expander   Device for increasing or restoring the dynamic range of a programme signal.

exponential movements   Animation where the speed of movements is proportional to the size of the field.

exposure   Process of subjecting a photo-sensitive surface to radiation, as in a photographic or video camera.

exposure meter   Instrument for determining the intensity of light incident on or reflected by a scene to be recorded by a photographic or video camera.

eye-lighting   Lighting a close shot to provide a small highlight reflection on the eyeball.

F

f   femto-, a prefix denoting a factor of 10^{-15}.

F   Farad, unit of capacitance.

f/-number   The relative aperture of a lens diaphragm, calculated by dividing its focal length by its diameter.

F & E   Type of video connector.

FACT   Federation Against Copyright Theft, in film and video (UK).

fade (1) The controlled gradual reduction or increase of a picture image or an audio or video signal. (2) Decrease in strength of a received signal because of transmission deficiencies.

fade down Gradual reduction of an audio signal or lamp intensity.

fade-in Gradual appearance of a picture, usually from a uniform black, or the increase of an audio signal.

fade-out Gradual disappearance of a picture, usually to uniform black, or the decrease of an audio signal.

fader Controller, usually a potentiometer, used to vary signal level or picture brightness to produce fade effects.

fader start Remote-equipment switch incorporated in a gain control fader.

fade shutter Two-bladed shutter in a motion picture camera, the opening of which can be altered during running to vary the exposure for fade and dissolve effects. In a printer, a variable aperture in the light beam for the same purpose. (= *fading shutter*)

fade up Gradual increase of an audio signal or lamp intensity.

fairings Acceleration or deceleration when movement in animation starts or stops or alters speed.

fall off Gradual reduction in illumination from centre of a screen to the edges and corners.

FAP *front axial projection.*

FCC (1) Federal Communications Commission (USA). (2) Frame Count Cueing. In motion picture printing, a control cue system based on electronic counting of the number of frames passing through the machine.

feedback, negative Condition where a fraction of the output of a system is fed back to the input in anti-phase, frequently used in electronic amplifiers. Input and output impedance, stability and linearity are affected.

feedback, positive Condition where the output of a system is fed back to the input in phase. There can be a build-up in amplitude leading to instability of the system, e.g. loudspeaker output fed back to the microphone input produces undesirable oscillations, 'howl-round' or *threshold howl.*

femto- A prefix denoting a factor of 10^{-15}.

ferrite A magnetic material of high *permeability*, used for the cores of high efficiency inductors in RF applications.

FET Field-Effect Transistor.

FFT Fast Fourier Transform.

FIAF Fédération Internationale des Archives du Film.

fibre optics System of light transmission along a fine flexible glass or plastic filament or bundle of filaments; used for the illumination of enclosed areas and for the cable distribution of pulse-coded light signals in telephony and television.

field In television, one complete scanning of the picture image from top to bottom; two interlaced fields are required to complete a frame.

field chart A plastic sheet, punched for register pegs, engraved with the exact area of the maximum field for a certain set-up.

field (data) A group of contiguous bytes forming a single piece of data.

field lens Large diameter supplementary lens used to increase light transfer in an optical system, especially in aerial image projection.

FIFO First In, First Out (Memory).

file (data) A named program or a collection of data.

fill To load a computer with data (usually a program) from disk or tape.

fill light Secondary lamp(s) used in the studio to fill shadows, usually opposing key lights.

film (1) A thin material carrying a layer of recording medium, especially light-sensitive photographic emulsion, on a transparent base. (2) Specifically, motion picture film, a comparatively narrow strip of such material, usually perforated along the edge(s) for transport and synchronisation. (3) Also used as a general term for the craft and techniques of cinematography.

film base The support upon which the light-sensitive photographic emulsion is coated.

film speed The sensitivity of a photographic film, as determined by standardised methods such as *ASA* or *DIN*.

filmstrip Compilation of individual still photographs, illustrations and/or text on a single length of film for projection purposes, often associated with lectures or demonstrations.

filter (colour) Transparent material modifying the colour of the transmitted light by selective absorption of certain wavelengths.

filter (electrical) System of components connected in series/parallel configurations to reject unwanted frequencies and pass the required ones. See *bandpass filter*.

filter (heat) Transparent material which transmits visible light but absorbs or reflects infra-red radiation (heat rays).

filter pack In motion picture printing with subtractive colour control, the group of colour and neutral density filters required to produce the correct exposure from a given negative.

fire rollers Pairs of small rollers through which film passes in a projector, intended to stop transmission of flame; also termed *fire-trap*.

firmware A combination of hardware and software to do some specific computer task, e.g. an operating system. The software part is usually in a non-volatile store such as *ROM*.

fishpole   Hand-held microphone boom, typically of 2 m length.

FKGT   Fernseh- und Kinotechnischen Gesellschaft, German film and television technical society.

flag   (1) A small opaque surface used in studio lighting to shade a particular area. (= *french flag*). (2) A *bistable* which can be set, reset or tested under program control.

flagging   Horizontal jitter at the top of a television picture, usually resulting from timing errors in the playback signal from a video tape recorder.

flanging   Audio effect similar to *phasing* but more intense.

flare   Scatter of light in a lens, which adds undesirable light to the dark areas of an image.

flash   A brief item interrupting a broadcast programme.

flash frame   In motion picture negative, a heavily exposed single frame producing a clear area in the corresponding positive print.

flat   Rectangular covered wooden frame used for building scenery.

Fletcher–Munson curves   Equal loudness graphs showing the frequency dependence of the human ear in relation to sound level for pure tones.

flicker   Random or regular short period variations of screen luminous intensity.

flies   Space above a stage or studio where scenery may be lifted.

flip   Momentarily switching from one projector to another in slide presentation.

flip chart   One of a sequence of large pages of text or diagrams usually presented on an easel. When each one has been used, it is 'flipped' over the top of the easel to disclose the next.

flip flop   In Europe, a circuit with two stable states, i.e. a *bistable*. In USA, a *monostable*.

FLLA   *fusible link logic array*.

float   (1) A movable part of a studio or stage set which can be readily lifted into or out of position. (2) In motion picture projection, a comparatively slow unsteadiness of the image in a vertical direction. (3) Movement of one image against another in multiple exposure rostrum camera work.

floating point arithmetic   A system of manipulation of numbers in calculators and computers that automatically takes the decimal point placings into account.

flood   *Luminaire* giving a wide spread of light.

flood track   A photographic sound track exposed to show the uniform maximum width of the image area.

flop-over   A visual effect in which the picture is shown reversed from left to right.

floppy disk Digital memory using flexible magnetic disks.

flow chart A graphical representation of the possible alternative routes available in a program.

fluid head A camera head for a tripod, with pan and tilt facilities made smooth by fluid damping.

fluting see *edgewave*.

flutter Variation in speed of a recording medium, occurring above 15 times per second.

flutter echo Effect caused by the repeated echo of sound between two facing parallel surfaces.

flux See *luminous flux* or *magnetic flux*.

flux density The intensity of a magnetic field. See *magnetic flux*.

flying erase Facility for erasing single video tracks in a helical-scan videotape recorder by a rotating head, rather than a fixed erase head, without loss of the audio or control tracks.

flying spot scanner Device for producing TV pictures from opaques or transparencies by scanning them with a spot of light, usually produced by a cathode ray tube. The reflected or transmitted light is collected by a photocell to produce the video signal; see *Figure F.1*.

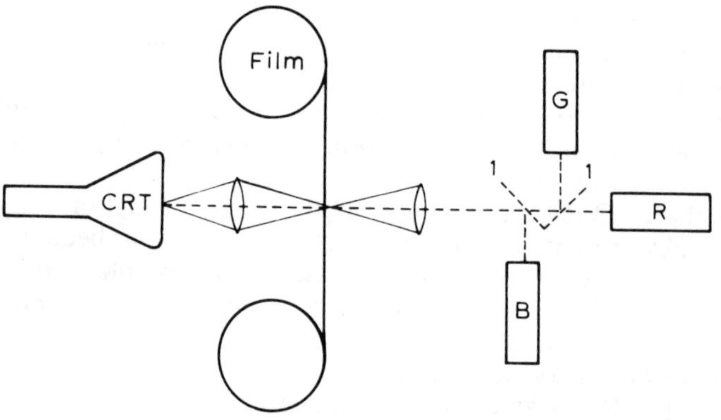

Figure F.1 Flying spot scanner. The raster on the CRT face is focused on the plane of the moving film. The light transmitted is divided into its RGB components by the dichroic mirrors 1, 1 and collected by three photo-multipliers to provide the signals

flyman Staff working in the fly tower on raising and lowering scenery, etc.

FM Frequency Modulation.

FO Fibre Optic (Link).

FOB Free On Board.

FOC Free Of Charge.

focal length   The distance from the rear *nodal point* of a lens to the point at which rays from an object at infinity form the most sharply defined image.

focal plane   The plane through the principal focus of a lens at right angles to its optical axis.

focus   To adjust the lens of a camera or projector, nearer to or further from the plane of the gate or aperture, in order that the image shall be sharply defined. Hence, the position at which the sharpest image is formed.

fog   Additional density in a processed photographic image caused by accidental exposure to light, radiation or a chemical fogging agent.

fog level   The minimum density of the unexposed area of processed film.

FOH   Front of house, as in FOH lighting.

foil   A sheet or roll of thin transparent or translucent material, either plain or carrying prepared images, as used on the platen of an *overhead projector.*

foldback   Cueing signal sent to musicians enabling them to hear a previously recorded performance, or that of another musician playing at the same time.

follow focus   A camera technique in motion picture and rostrum photography of compensating for varying subject distance to maintain image focus.

follow spot   Powerful spotlight used to follow the movement of an artiste.

font   Originally, a set of printers' type of uniform style and size ('fount'); by extension, a uniform style of alpha-numeric characters available for selection in a video graphics system.

foot   Measure of motion picture film length representing a specific number of frames: 16 for 35mm, 40 for 16mm and 72 for Super-8.

footage numbers   See *edge numbers.*

footprint   The area of coverage of a satellite transmitter on the surface of the Earth.

forced development   Motion picture processing using increased time or temperature to compensate for under-exposure of the original film in the camera.

form feed   In word-processing, a character which causes the print or display position to move to the start of the next page, the length of the page having been defined; often abbreviated **FF**.

format   (1) The style or method of presentation of an image or its physical dimensions. (2) The configuration of information for display, transmission, storage or retrieval.

FORTRAN   FORmular TRANslator. A problem oriented level computer programming language.

4:2:2   Refers to the ratio of digital TV sampling frequencies for the *luminance* and *chrominance* signals specified in *CCIR Rec. 601*.

FP   Front Projection.

FPGA   Field Programmable Gate Array.

FPLA   Field Programmable Logic Array.

FPLS   Field Programmable Logic Sequencer.

fps   Frames per second.

FR3   France Région 3, the third French state broadcast TV network.

frame   (1) An individual picture image on motion picture film. (2) In video, the picture formed by a pair of interlaced fields. (3) In computer graphics, one *refresh* of a display image.

frame grab   In computer graphics, temporary storage of a video frame for later manipulation by a graphics input device.

frame counter   In motion picture equipment, a counter registering the number of frames of film passing through it.

frame line   The space between consecutive frames on film.

frame rate   For sound film a rate of 24 frames per second has been standard since 1927 but a rate of 30 fps has been proposed for improved quality of presentation; in television and video 30 fps is standard for *NTSC* systems and 25 fps for *PAL* and *SECAM*.

frame store   A storage system capable of storing a complete frame of video information in digital form. Used for television standards conversion and sophisticated special effects units.

framing   Adjusting the film image in the projector gate aperture so that the projected image appears centred in the vertical dimension of the screen.

free field   Sound field with no reflective surfaces.

free head   A recording head without electronics, provided on an audio cassette recorder for use with tape-slide control equipment.

freeze frame   A visual effect in which a single picture from a moving scene is repeated to hold the action stationary for as long as is required. Also called *hold frame*.

freeze grab   Facility in a TV field or frame store to monitor a continuous input signal and select a field or frame at a predetermined rate, reproducing this as a still frame until updated by the next.

french brace   A strut used to support scenery.

frequency   The number of times that a periodic function or oscillation repeats itself in a specified time, usually one second. Hertz (oscillations per second) is the preferred unit; cycles-per-second is deprecated.

frequency modulation   A method of passing information by altering the frequency of the carrier signal.

frequency response   The relationship between the *gain* of a device
and frequency.

Fresnel lens   A lens formed by concentric stepped rings, each of
which is a section of a convex surface. Often used as a condenser
lens in an overhead projector or spotlight; see *Figure F.2*.

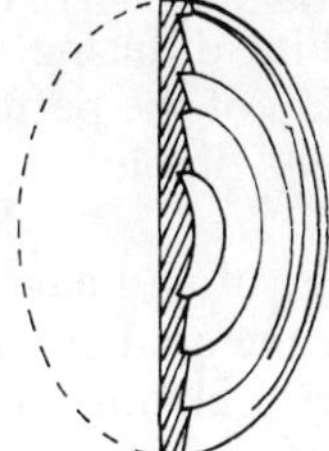

Figure F.2 Fresnel lens

fringing   (1) False colouration around an image arising from a
defective or maladjusted optical system. (2) The effect of using a
full track test tape on a multi-track tape machine and so giving
an exaggerated low-frequency response.

front axial projection   A trick effects shot, in which the background to the action is provided by projecting an image along
the axis of the camera lens on to a highly directional beaded
reflecting screen. Also known as *reflex projection*.

front end   In motion picture production, front end operations are
those concerned with original photography, sound recording,
editing and preparatory work leading up to the first presentation.

front porch   In a television waveform, the very brief 'black'
period between the end of picture information and the start of
the horizontal sync pulse; see *Figure B.1*.

front projection   The presentation of a picture to be viewed from
the same side of the screen as the projector.

full bleed   Full-page illustration which runs off the page at all or
some of its edges.

fuse   An intentional weak link in an electrical circuit designed to
fail if the rated current is exceeded.

fusible link PROM   A read-only memory device which is programmed by electrically rupturing internal links.

fusion point   In computer graphics, the point at which the rate of
refresh redrawing the display makes it appear steady.

FVPG   Film and Video Press Group (UK).

FX   Effects.

g gram.

G *gauss*.

G giga-, a prefix denoting a factor of 10^9.

Ga-As Gallium Arsenide.

gaffer Senior lighting electrician in a film or TV studio.

gain (1) Ratio of amplification or attenuation, usually expressed in decibels. (2) The measure of the reflectivity of a surface compared with a perfect matt white diffusing surface under identical conditions. (3) The measure of the performance of an aerial system compared with a simple dipole.

gallery Studio control room.

gamma Measure of the tonal contrast of an image reproduction process. (1) In photographic terms, the gradient of the straight-line portion of the characteristic curve relating log-exposure and density, expressed as the tangent of the angle which it makes with the base line. (2) In television, the overall gamma is the relation between the logs of the luminance increment on the receiver screen and the corresponding increment in the original scene. For a camera it is the slope of the curve relating the logarithms of the incident light and the output voltage; similarly for a display tube, the input voltage and the light output. See *Figure G.1.*

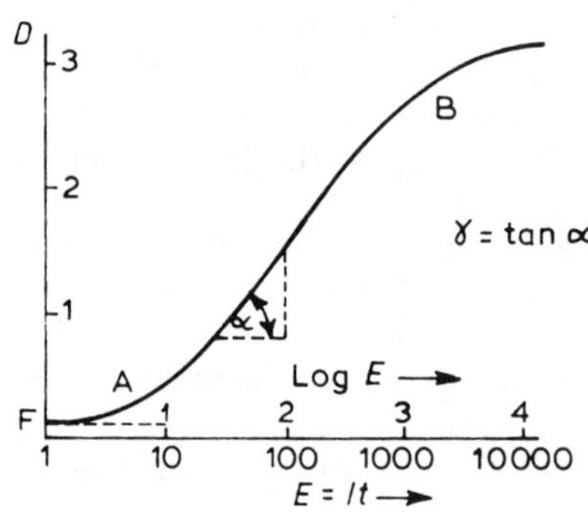

Figure G.1 Gamma. In photographic reproduction, gamma represents the slope of the straight-line portion of the characteristic curve

gamma correction In television, non-linear amplification applied through the transmission chain to obtain satisfactory reproduction within its limitation.

ganged Optional mechanical locking together of controls so that movement of one causes movement of the others.

gap The separation between the poles of a magnetic recording head at right angles to the direction of motion of the tape. The width of the gap determines the highest frequency which can be handled. (= *head gap*).

garbage collection see *housecleaning*.

gas-discharge Usually refers to light sources in which an electrical discharge takes place in a rare gas within an envelope to provide the illumination.

gate (1) In electronics, an element of digital logic, e.g. *AND, OR, NAND, NOR, EXCLUSIVE-OR*. (2) Switching a signal electronically. (3) In photographic equipment, the aperture at which motion picture film is exposed or projected. Also the assembly that supports and positions the slide in a slide projector.

gate temperature In an optical projection system, the temperature reached at the gate aperture.

gate turn-off switch A special type of *thyristor* which can be turned off under gate control even when anode current is flowing.

gauge The width of film or tape, usually expressed in millimetres.

gauss Unit of magnetic flux density.

gauze Transparent or translucent material for special effects.

geared head A camera head for a tripod, with geared pan and tilt features.

gel Filter for lighting, which changes the colour or quality of light.

generation loss Degradation of video picture quality caused by successive stages of transfer or dubbing.

geneva movement A mechanical intermittent motion device based on a maltese cross shape, interacting with a pin rotating on a disc.

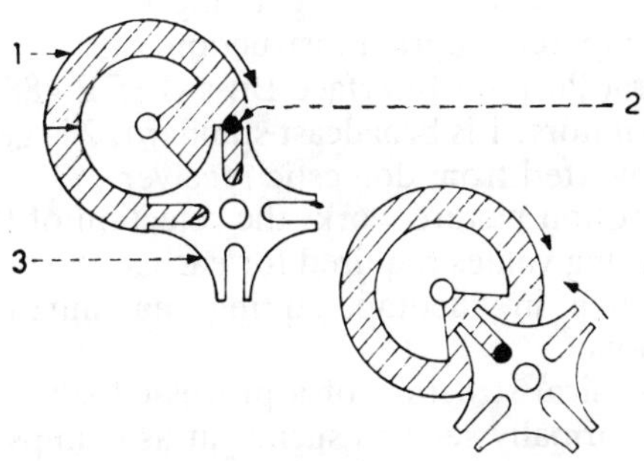

Figure G.2 Geneva movement. The rotating disc 1 carries a pin 2 whose engagement in the slots of the maltese cross 3 produces the intermittent movement of the latter

genlock Means of locking local sync pulse generator (SPG) to incoming or pre-recorded video signals to provide full synchronism.

GEO Geosynchronous Equatorial Orbit, in which a satellite is effectively geo-stationary relative to a point on the Earth.

getter Substance used to remove, by chemical combination, unwanted gases in valves, cathode ray tubes, etc.

ghosting Shadow or weak double images appearing in a picture. In television can be caused by unwanted reflections in a cable system or from physical objects in a radiated signal. In film projection it can result from an out-of-phase shutter.

GHz Gigahertz; see *hertz*.

giga- A prefix denoting a factor of 10^9.

GIGO Garbage In, Garbage Out, a colloquial maxim in computing.

gliding frequency (1) An audio response test using a continuous slowly changing frequency, usually from 20 Hz to 20 kHz. (2) A form of tape-slide control using *frequency modulation*.

glitch An unwanted transient causing a brief disturbance, as of the picture in video, or misoperation in digital devices. Hence a defect in A–D and D–A converters giving such transients.

global beam A transmission covering all the surface of the Earth visible from a satellite.

GMT Greenwich Mean Time.

GMX Graphics model exchange.

gobo A portable opaque shield, usually stand mounted, to act as a light shield between a lighting unit and camera lens. In the theatre gobos are fitted to *luminaires*.

golfball A type of printer or typewriter with the printing characters embossed on a small sphere.

GOTO Go To. An instruction signifying an immediate jump to a different computer program instruction.

GPIB General Purpose Interface Buss (IEEE 488).

grades In monitors, 1 is broadcast standard, 2 is general purpose and 3 is converted from domestic receiver.

grading In motion picture work, the selection of the colour and density printing values required for each scene of an assembled roll of negative; in videotape editing, matching colour balance between shots.

grain The physical structure of a processed silver photographic image; not normally seen as such, but as clumps of grains, for which the term graininess is more appropriate.

graphic equaliser A multi-section frequency selective filter providing instant adjustment of the 'shape' of the frequency response characteristic.

graphics   (1) Artwork, captions, lettering, photographs, etc. used in programmes. (2) Output presented in graphical form on a visual display unit.

graphic tablet   Device for input of details of graphical information to a computer; see also *data tablet*.

graticule   A grid or grating existing as a physical structure or as a deposited pattern on a glass or plastic support, used particularly for the measurement or location of an image.

gravure coating   A method of coating magnetic tape similar to the gravure printing process.

green print   A processed motion picture film having an excess humidity in the emulsion, which may cause unsteadiness or damage in projection.

grey scale   A test chart with incremental brightness steps from black through grey to white; the steps may conform to a logarithmic relationship.

grid   (1) Uniformly spaced points or lines in two or three dimensions within which an object may be defined. (2) The current control electrode of a valve, cathode ray tube, or camera tube. (3) Framework above studio floor for suspending scenery and lights.

grille   Test waveform for converging colour monitors; see *cross hatch*.

grip   Member of film crew transporting equipment or pushing a camera.

ground   An electrical zero-voltage reference, i.e. at earth potential or connected to earth.

grounding   Electrically connecting to earth.

ground loop   A fault condition affecting signals passed between units where circulating currents in multiple ground paths introduce hum.

ground row   Lighting units on a stage or studio floor for lighting cycloramas or scenery.

ground station   Transmitting or receiving station on Earth forming part of a satellite communications system.

group   (1) The ganging together of a selection of circuits on a lighting desk or audio mixer. (2) Controls on a mixing console or lighting desk simultaneously affecting a selected set of channels.

group delay   Effect in transmission system where some frequencies travel faster than others–not the same as *phase shift*.

G-spool   *Spool* holding 5 minutes of videotape, used for commercials.

GTO   Gate Turn Off Device. Similar to a *thyristor* but a GTO turns off when gate current ceases to flow instead of latching until anode current falls.

guard band Unrecorded region between tracks on magnetic tape in order to avoid *crosstalk*. Also unused radio frequency spectrum between radio channels.

guide In magnetic recorders, indented pillars or rollers used to define the tape path.

guide path Posts and guides that shape the tape path in audio and video recorders.

guide track A speech track recorded where background noise is high, to serve as a guide for the actor to do the speech again in a studio.

gun microphone A very directional microphone, also called a *rifle microphone*.

gutters Dark bands between the image areas of a compartmented multi-screen; see *egg boxing*.

G/V General View, an establishing shot in film or video.

H

h hecto-, a prefix denoting a factor of one hundred, 10^2.

H Henry, the unit of inductance.

hack A journalist.

halation Unwanted exposure surrounding the photographic image of a bright object, caused by light scattered within the emulsion layer or reflected from the base of the film.

half duplex Mode of data transmission in which each station can send to the other, but only in one direction at a time.

half frame One having dimensions smaller than the standard (35mm) frame, i.e. 24×18 mm instead of 24×36 mm for slides or 21×8 mm rather than 21×16 mm for cinematography.

half-tone Tone differention by a series of ruled, etched or pigmented lines or vignetted dots evenly spaced at specific intervals.

half track A magnetic recording with two separate tracks on the same tape. Half track heads are narrower than stereo heads and have a wider *guard band* between them.

half wave Term used for circuit operating on the positive or negative halves of an alternating current, but not on both halves.

Hall effect A method of sensing a magnetic field by passing a current through a semiconductor and detecting a voltage produced. It is used in some keyboards and proximity detectors.

hand control A manual control for operating an automatic presentation device, remote from that device.

hand-held   Use of a camera without a tripod mount.

hand-shake   A two-way interchange between a computer and a peripheral, e.g. to enable a printer to signal if it is ready or not ready to receive data.

hard copy   A printed, thermally or electrically produced print-out as opposed to the image on a visual display unit.

hard cut   In slide projection, a very fast *cut*, as rapid as lamp inertia will permit.

hard disk   A digital magnetic data store comprising a non-magnetic rigid disk coated with a magnetic material.

hard edge   A sharply defined boundary to a picture image area, as in a *matte* or *wipe*.

hardware   The physical components, integrated circuits, etc. which, with the software program, together make a computer system. The term is also used to refer to picture and sound equipment used for presentation, as opposed to the programme software.

harmonic   Frequencies being a whole multiple of the fundamental frequency.

harmonica connector   A multi-pole strip connector resembling a harmonica from the holes in its construction. Also known as a *chocolate block*.

harmonic distortion   The unwanted generation of harmonics.

harmoniser   Electronic *pitch changer* to raise or lower pitch without altering the time aspects of a programme.

hash marks   Unofficial *cue marks* scratched on a film release print.

Hawkeye   Trade name for a *camcorder* system by RCA.

HBST   High Band Saticon Trinicon camera tube.

HD-NTSC   Trade mark of proposed US high-definition TV system with 828 pixel lines and AR 14:9 but providing signals compatible with existing *NTSC* channels and receivers.

HDTV   High Definition Television.

head   (1) General term for the essential mechanism of a recorder or reproducer, as in Projector Head. (2) In magnetic recording, an electromagnetic transducer converting magnetism to electrical signals and vice versa. (3) In photographic sound reproduction, the sound head converts the light passing through the sound track on the film into an electrical signal. (4) The adjustable mounting for a camera on its tripod. (5) The beginning of a roll of film or tape.

head banding   see *banding*.

head end   Originating point on *CATV* or aerial system, usually with an amplifier.

head demagnetiser   A device producing an alternating magnetic

flux at a probe. It is used for demagnetising tape heads and
guides.

head out   Film or tape wound on a reel so that it is at the start,
ready for immediate projection.

headphone(s)   Small sound reproducers used over, or in, the ears,
usually excluding extraneous sounds.

head room   Dynamic range available above working level, before
overload distortion occurs.

headset   Headphone/microphone combination.

head wheel   see *drum (2)*.

heat filter   A filter which by absorbing or reflecting infra-red
radiation while transmitting visible light, reduces the heating
effect of the beam.

heat sink   A device, usually metal, to dissipate unwanted heat by
radiation and/or convection, e.g. a finned plate to cool a power
transistor.

hecto-   A prefix denoting a factor of 100, 10^2.

helical scan   System of video tape scanning in which the tape
forms a helix around a rotating drum containing the magnetic
head or heads; see *Figure H.1*.

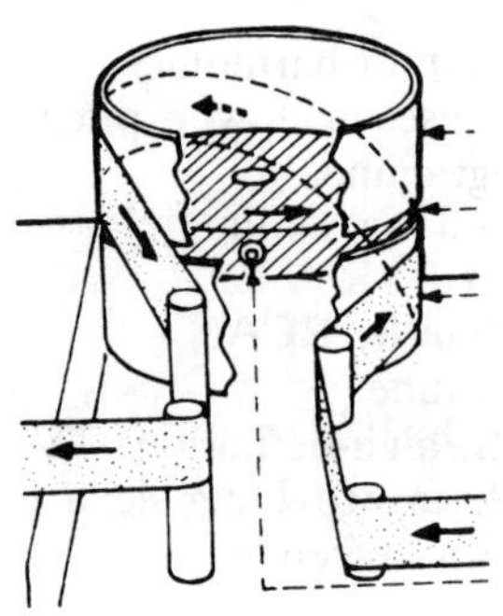
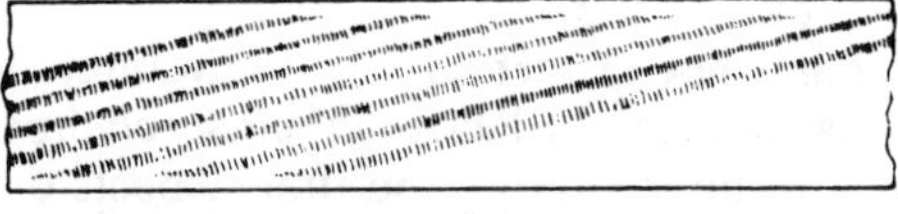

Figure H.1 Helical scan. In helical scan, the tracks are long diagonals on
narrow-gauge tape.

hertz   Unit of frequency of oscillation; one cycle per second.

heterodyne   The production of a beat frequency from the mixing
of two alternating signals of which it is the combination of the
sum and difference frequencies. The technique is used in radio
and TV reception.

hexadecimal   Method of expressing the 16 possible states of 4
binary *bits*. These states are expressed as a single character
0,1,2,3,4,5,6,7,8,9,A,B,C,D,E or F. Eight-bit *bytes* are dealt
with as two four-bit quantities, e.g. A7 hex = 10100111 binary.

hex code   Microcomputer program in hexadecimal.

HI Arc   High Intensity Arc, a carbon arc lamp operating at a high current density.

hidden lines   In computer graphics, line segments which should not be visible to the viewer of a three-dimensional display item because they are behind other parts of the same or other items. Similarly *hidden objects, hidden surfaces;* see *Figure H.2.*

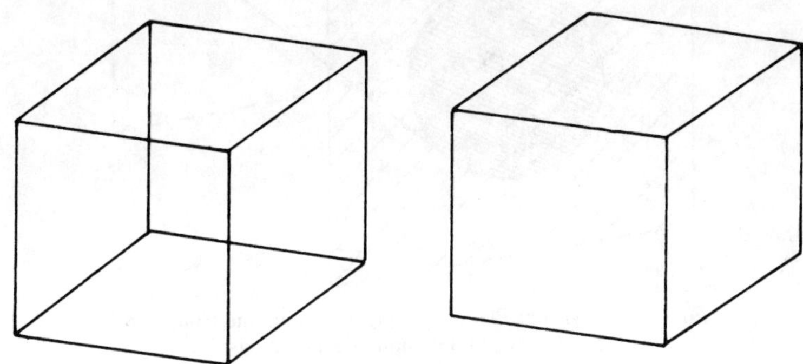

Figure H.2   Hidden Lines. In computer graphics, hidden lines must be removed to represent a three-dimensional object

high band   A videotape recording system producing broadcast quality pictures, such as the 1 inch format and ¾ inch *BVU*.

high frequency   In radio frequency terms, 3 MHz to 30 MHz.

high key   Characteristic of a scene in which the majority of tones in the subject or its reproduction are at the light end of the scale.

high level language   A set of computer programming words, etc., of a powerful nature. The language should be as similar as possible to English. Some examples are ADA, ALGOL, APL, BASIC, COBOL, COMAL, CORAL, FORTH, FORTRAN, LISP, PASCAL, PILOT, etc.

highlight   The brightest part of a scene or its reproduction.

high-pass filter   An electrical filter which passes high frequencies and attenuates low frequencies.

hither plane   In a three-dimensional graphics display, the front *clipping* plane that defines a finite view volume; see *Figure H.3.*

Hi-Vision   Name adopted for the Japanese *HDTV* system, specifically using 1125 lines at 60 fields with 16:9 aspect ratio.

HMI   Trade name for a double-ended *metal halide lamp* with a colour temperature of 5600K.

hold frame   See *freeze frame.*

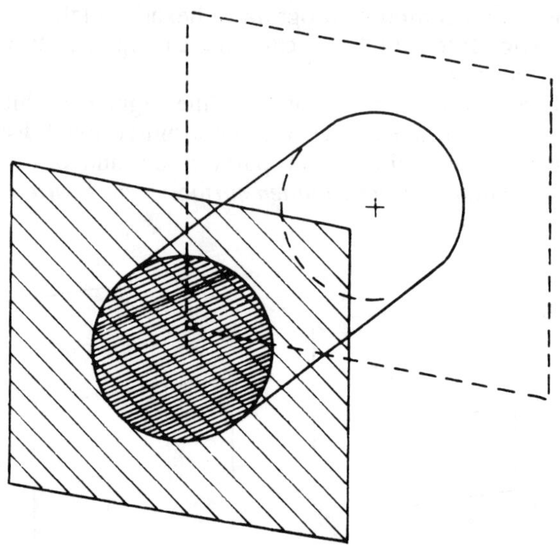

Figure H.3 Hither Plane. The plane limiting the front of the
viewed space in a computer graphics display

holography The recording and representation of three-dimensional objects by interference patterns formed between a beam of coherent light and its reflection from the subject.

home At the base or starting position. Rotary slide projector magazines with a zero position usually 'home' to zero.

homing Term used with automatic slide projectors for returning the magazines to the start or 'home' position. Circular ones may be homed by the shortest route.

horn Acoustic portion of a sound reinforcement loudspeaker in which the cross-section increases in area towards the output end.

horse A simple support with a spindle for a roll of film wound on a core.

hot shoe A still-camera accessory socket with integral shutter-contact connection for flash.

hot spot An area of excessive local brightness, particularly as part of a projected image or illuminated scene.

housecleaning Freeing component memory, e.g. *string* space, that is no longer pointed at by valid pointers. Also termed *garbage collection*.

housekeeping Looking after the details of automatic presentation in a cinema, such as operating *houselights*, *tabs*, etc.

houselights Lights illuminating the audience seating area or general lighting in a studio.

hub The cylindrical centre of a film or tape *spool*; sometimes used to mean a *core*.

hue The attribute by which a colour may be identified within the visible spectrum. A hue control may be found on some television receivers.

hum An audible continuous sound of low frequency, often corresponding to mains frequency of 50 or 60 Hz, or its harmonics.

hum bars Horizontal slow moving bars on a television picture caused by injection of unwanted mains hum into the video signal.

hunting (1) A cyclic error in a servo system. (2) In video tape reproduction, a low-frequency instability of picture or sound caused by cyclic variations in tape or film transport velocity.

hyper-cardioid microphone Microphone with directional sensitivity pattern between that of a cardioid and figure-of-eight.

hypersensitising Increasing the sensitivity of a photographic material before camera exposure, by chemical means or by brief controlled pre-exposure to light.

hypo Sodium thiosulphate used as a photographic fixer.

Hz Hertz. Frequency in cycles per second.

I

IAC Institute of Amateur Cinematographers (UK).

IATSE International Alliance of Theatrical Stage Employees, and Moving Picture Machine Operators of the United States and Canada.

IBA Independent Broadcasting Authority. Formerly ITA (UK).

IBC International Broadcasting Convention.

IC Integrated Circuit.

ICE Insertion Communication Equipment. BBC communication system using a line in the field blanking period.

ICE In Circuit Emulation. A microprocessor development tool.

icon In computers and in equipment labelling, a symbol clearly representing purpose or function.

IEC International Electrotechnical Commission.

IEE The Institution of Electrical Engineers (UK).

IEEE The Institute of Electrical and Electronics Engineers, Inc (USA).

IEEE-488 A parallel format communications *bus* and *protocol* often used for controlling laboratory instruments. Also known as IEC-625, HP-1B and GP-1B.

IERE Institution of Electronic and Radio Engineers (UK).

IF Intermediate Frequency.

IFPP International Federation of Photogram (and Video-gram) Producers.

IGFET Insulated Gate Field-Effect Transistor.

igniter Device to strike a xenon or mercury arc.

ILD Injection Laser Diode.

ILR Independent Local Radio (UK).

IM Intermodulation.

image bit map Digital representation of a graphics display image as a pattern of bits, each bit representing one or more *pixels.*

image enhancer Device for improving video picture quality, especially in apparent resolution or tonal reproduction.

image intensifier Device which enables a camera to operate under extremely low light levels, typically starlight.

IMAX Wide-screen motion picture system using 70 mm film running horizontally with a 15-perforation interval, giving a frame size of 70×46 mm.

immediate A computer command which will be acted upon at once in real time, rather than at run time.

impairment scale Scale of subjective assessment of the loss of quality observed in the reproduction of sound or picture, both film and video. Scale 5 is negligible loss, Scale 1 unacceptable.

impedance Resistance to alternating current or motion.

impulse A mechanical force or electrical signal operating for a very short time.

in-betweening In animation, generating the movement of a figure between two specified positions.

infra-red Radiation at wavelengths extending from just beyond the deep red end of the visible spectrum, above 800 nm. up to UHF waves. Generally discernible as heat.

initialise To get into its correct starting state.

ink-jet plotter A *plotter* or printer using a computer-controlled electrostatically steered stream of ink droplets.

inkie Incandescent light using tungsten or tungsten-halogen lamps, as distinct from carbon arc or metal-halide sources. (= *inky*)

inking In computer graphics, using an input to sketch freehand with a stylus.

inlay Video effect where one TV image is inserted into part of another.

in-line tube   Type of colour TV tube with electron guns in-line rather than in a triangular formation.

input   The data or signals entering a computer or microcomputer; more generally, the signal applied to any electronic device or system.

insert   In video editing, any material replacing a section within an existing recording. In film, it is usually a scene photographed separately and added during editing, such as a close-up shot of a letter or a newspaper.

in-store video   Public video display for point-of-sale information, sometimes *interactive*.

instruction   Command forming part of a computer program. A sequence of instructions causes the machine to execute its task.

insulation displacement connector   A form of connecting system where wires are terminated without removing their insulation. The connector displaces the insulation to obtain a good electrical connection.

integer   (1) A whole number. (2) In computer practice, a type of *variable* consisting of a whole number with no fractional or decimal component, e.g. MONTH% = 11.

integrator   An electrical circuit which sums its input signals with regard to time.

intensification   The process of increasing the density of a previously developed photographic image.

interactive   A video or computer program in which there is a two-way interplay between the viewer or operator and the machine.

intercom   A two-way voice communication channel between locations.

intercutting   A method of editing in which varied shots of the same subject are edited together.

inter-dupe   A duplicate colour negative film derived by printing from an inter-positive.

interface   A circuit that converts the output of one system or sub-system to a form suitable for the next system or sub-system.

interlace   Television scanning system in which the sets of lines in each pair of fields are displaced vertically so as to alternate with each other; see *Figure I.1*.

interlock   To synchronise and hold in lock.

intermediate   General term for colour film master positives and duplicate negatives printed on a film stock having integral colour masking.

intermediate level   A dimmer setting, usually preset, which lies between off and full up.

intermittent motion   The transport system in a camera, step

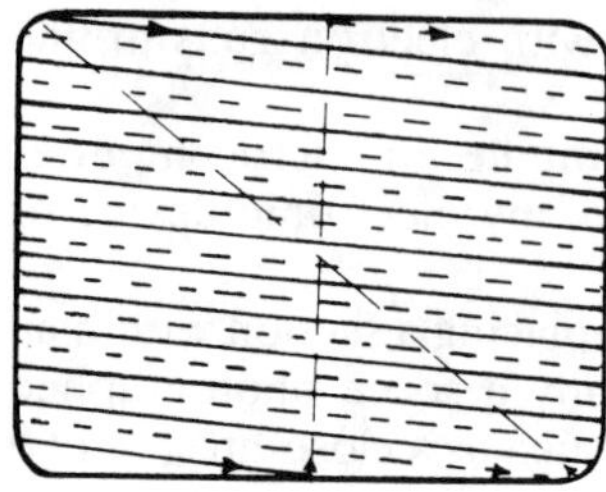

Figure I.1 Interlace. All the odd-numbered lines of the first field (solid) are scanned and are followed by the even-numbered lines of the second field (dashed) to complete each frame

printer or projector in which each frame in succession is moved into position and held stationary for the period of exposure or presentation.

intermodulation   The generation of unwanted sum and difference signals when two or more frequencies pass through a non-linear system.

inter-negative   A duplicate colour negative film, especially one prepared directly from an original reversal film exposed in the camera.

inter-positive   A colour master positive film printed directly from an original colour negative on to intermediate stock.

interpreter   A computer program which translates a high-level language line by line, to produce executable code at run time.

interrupt   A method of gaining the immediate attention of a computer. The computer is programmed to respond in a particular way before resuming its previous task.

inverse   In data presentation, the exchange of black and white to produce black lettering and lines on a white background on a monitor and white on black in a printer.

I/O   Input/Output, system of a microcomputer.

IPPA   Independent Programme Producers Association (UK).

ips   (1) Inches per second. (2) Instructions per second.

iris   Adjustable *diaphragm* to regulate the amount of light passing through it, as in the eye or in a camera lens.

iris wipe   A *wipe effect* in the form of a circle, expanding for *iris in* and diminishing for *iris out*.

I-signal   In the *NTSC* colour system, represents the *chrominance* on the orange-cyan axis.

ISO   International Organization for Standardization.

isochromatic   Substantially sensitive to colours in their relative visual intensity.

isolating transformer   A transformer with separate primary and secondary windings to provide isolation and safety, usually from the mains supply.

isopropyl alcohol   An organic liquid solvent, used for cleaning magnetic heads and tape paths.

IT   Information Technology.

ITA   (1) Before 1972, Independent Television Authority. (2) International Tape Association (audio, including disc).

ITCA   Former Independent Television Contractors Association (UK), now ITV Association.

ITN   Independent Television News (UK).

ITS   Insertion Test Signals. Test waveforms inserted in field blanking enabling picture quality to be quantitatively tested.

ITU   International Telecommunications Union.

ITV   Independent Television (UK).

ITVA   International Television Association.

ITV Association   Independent TV Association (UK), formerly ITCA.

IVCA   International Visual Communications Association, an affiliation of *ITVA* (UK) with the former *BISFA*.

J

J   Joule, unit of energy, work or heat.

jack   A cable-mounted plug of circular cross section usually with two or three in-line contacts, often used for audio frequency signals.

jack field   Strips containing multiple jack sockets.

jamb-sync   A regeneration of *time-code*, particularly when editing videotape, to ensure the numbers are in ascending sequence throughout the tape.

JCII   Japan Camera Inspection Institute.

jelly   A colour filter or diffuser placed in front of a lighting source. (= *gel*).

JFET   Junction Field-Effect Transistor.

jib   The arm of a camera crane.

JICCAR   Joint Industry Committee for Cable Audience Research (UK).

jitter   (1) Small irregular unsteadiness of the film in camera or projector. (2) Subjective effect of flutter in a reproduced video image. (3) Small irregular fluctuations of the timing of a pulse or waveform edge. (4) Jerky movement due to faulty animation.

jog   Facility on film or video equipment to move backwards or forwards by a small amount.

join   A splice or edit between two shots.

joystick   A control lever capable of movement in two planes, for example to control a *cursor.*

jumbo   A wide roll of magnetic tape or photographic film produced by the coating machine before slitting into strips of the final widths.

jumbo slide   A transparency of larger size than the conventional 24 × 36 mm slide (50 × 50 mm mount), usually 55 × 55 mm.

jump   A term for a leap to a new place in the execution of a microprocessor program. See also *conditional branch.*

jump cut   In film and video editing, the deletion of a section within the continuous action of a scene.

K

k   (1) kilo. A prefix denoting a factor of one thousand, 10^3. (2) Abbreviation in computer practice for 1024 *bytes*, as in 64k memory.

K   *Kelvin.*

kansas city   A 'standard' for recording computer data or programs on cassette tape.

K-band   Frequency band 10.9–36 GHz for satellite use.

kelvin   Scientific unit of temperature, equal to 1 degree Celsius, on a scale the zero of which corresponds to −273° C.

kerning   In data presentation, controlling the horizontal distance between characters to improve appearance.

keyboard   A man–machine interface where information is entered on buttons or keys, as used in musical instruments and computers.

keyer   A vision mixer effect that cuts a hole in the main picture the outline of which is determined by another camera output. Thus captions may be keyed into a picture. (cf. *chromakey*).

key frame animation   Graphics of the principal poses 'drawn' on a cathode ray tube display and stored in a computer for animation use.

key numbers   See *edge numbers.*

key pad   A small keyboard or part of a keyboard, often used for numeric entry; it has numerals 0–9, and may have + − and other keys.

key light   The principal light illuminating the main subject in a scene.

keystone distortion   Image distortion in an optical or video system causing a rectangle to appear with both vertical sides converging more or less equally (cf. *trapezium* distortion); see *Figure K.1.*

kHz   kilohertz.

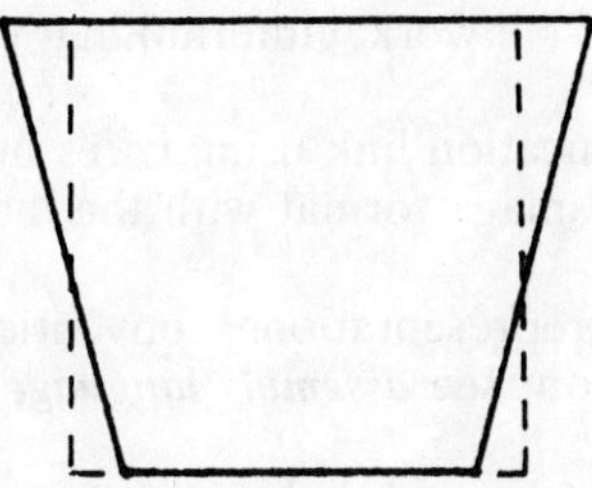

Figure K.1 Keystone distortion

kilo Strictly a prefix denoting a factor of one thousand, but often misused for kilogram, one thousand grams.

kicker A spot lamp used to give a high-light effect in back lighting.

kinescope (1) American term for a film recording of television (= *telerecording*). (2) American term for a cathode ray tube.

kludge An untidy modification or adaptation, usually applied to hardware.

knee compression Controlled non-linearity of tonal reproduction in a video system, enabling it to handle a wider range of subject contrast.

Kodatrace A matt-surfaced translucent plastic sheet material. (Trade mark).

KSR Keyboard Send/Receive, a teleprinter term.

Ku-band Frequency band 10.9–11.7 GHz for *DBS* transmission.

kVA kilo-Volt Amperes.

kW kilowatts.

L

l litre.

L Lambert. Unit of surface brightness.

lace To thread a motion picture projector or tape recorder.

lacquering Coating the surface of motion picture film with a transparent layer to reduce or eliminate the visibility of scratches or abrasions.

LAD Laboratory Aim Density, a value specified in the control of processing intermediate film.

lamp-failure detector Device to detect and report projector lamp failure. See *ALC*.

LAN Local Area Network, interlinking computers within a limited area.

land line Communication link using wires or cable, not radio.

landscape Picture image format with the major dimension horizontal.

language A set of representations, conventions and rules used to convey information; see *assembly language, high level*, and *low level language*.

lap dissolve Short for overlapping *dissolve*.

laser Light Amplification by Stimulated Emission of Radiation. A device for generating a beam of coherent monochromatic radiation.

laser printer A very high-quality data printer based on a combination of laser scanning and *xerography*.

Laservision A videodisc system employing a laser beam to read an encoded reflective grooveless disk. (Trade name).

latch In electronics, a *bistable* used to hold a state from one sampling instant to the next.

latching In a *thyristor* or *triac*, the device being held in its conducting condition by the continued presence of anode current.

latensification Intensification of an under-exposed image by controlled fogging by a light source before development.

latent image The recorded invisible image contained in exposed but unprocessed photographic film.

latitude The acceptable exposure range possible with a given photographic stock.

lavalier A neck-hung microphone.

layout A pencil drawing showing the main features of the background for a scene for animation.

L-band Frequencies about 1.6 GHz used for satellite down-link.

LCCC Leadless Ceramic Chip Carrier.

LCD Liquid Crystal Display. A reflective or transmissive alpha-numeric display of very low power consumption.

LCT arc Low Colour Temperature arc lamp.

leader (1) A length of film or tape at the beginning and end of a reel providing identification, information and protection. (2) A length of coloured non-magnetic tape attached to the beginning and end of some magnetic recording tapes.

leading In data presentation, controlling the vertical distance between rows of characters, on the analogy of type-setting.

LED *light-emitting diode.*

leisure time programming Programming a tape-slide or multivision show at the operator's own pace, one cue at a time, rather than in *real time*.

lens   An optical device of one or more elements in an illuminating or image-forming system, such as the objective of a camera or projector.

lens aperture   The effective opening of a lens system. It may be expressed as a fraction (f/- number, the ratio of focal length to physical opening) or as a measured factor of transmission, *T number*.

lens cap   A light- and dust-tight cover to protect the elements of a lens when not in use.

lens hood   An extension of the barrel of a lens, in solid or bellows form, used to restrict stray light entering the lens.

lens mount   That part of the body of the lens housing which engages in the equipment to which it is attached. It may have a screw thread of standard size (e.g. 'C' mounting) or a bayonet or other fixing.

lens turret   A rotatable mounting carrying a choice of two or more lenses, any one of which can rapidly be brought into use; see *Figure L.1*.

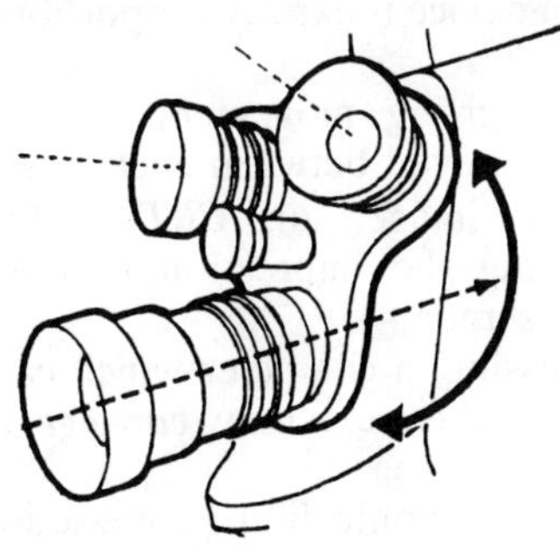

Figure L.1 Lens turret

lenticular screen   A projection screen the surface of which is composed of small lenses or ribs to improve image brightness by increasing the amount of light reflected in specific directions.

Leslie speaker   A loudspeaker on a rotating baffle or surrounded by a rotating cylinder with a slot in it. (Originally a trade name).

level   (1) The amplitude of an audio or video signal. (2) Sound volume, measured in decibels. (3) The position of a *cel* in a stack of cels, that nearest the background being level 1.

library   A store of picture and/or sound information available for reference and further use.

library material   Recordings intended for and held in a library.

library music, -shot, -slide   Music recording, picture scene or slide derived from a library, not original recordings made for the production concerned. Library music may be mood music specially written for use in audio and audio-visual productions.

LIFO   Last In, First Out (memory).

lift   See *pedestal*.

light box   An illuminated panel against which film or slides may be examined; also used for tracing in animation.

light pen   A stylus with an optical sensor to allow interactive use with a visual display unit screen, as in graphics.

light-emitting diode   A semi-conductor junction which when conducting emits visible light, used for indicators and alphanumeric displays, or infra-red radiation, used for position sensing and remote control.

lighting grid   Lighting gantry above a studio floor. See also *grid*.

lighting plot   A plan showing the number and position of *luminaires*, also the programme of lighting cues and states for their use.

lighting rig   The arrangement of *luminaires* and their accessories.

limes   Jargon for *FOH follow spots*.

limiter   Device restricting the maximum level of a signal.

limiting   Restricting the amplitude of a signal.

linear   Where the output of a device is exactly proportional to the input.

linear movement   Animation giving movement at a constant speed. The same distance is moved between each frame.

line crawl   Visual defect of images on *CRTs* with short-persistence phosphors, giving the impression that lines are moving up or down on the screen.

line feed (LF)   In word processing, a character which causes the print or display to move to the next line; *carriage return* is needed to move to the start.

line film   Ortho-sensitive photographic film processed to high contrast, for maximum separation between black and white.

line follower   In computer graphics, an input device which detects and traces lines in *vector* format.

line level   Audio signals peaking above 10 dBm (0 dBm = 1 milliwatt in a 600 ohm circuit).

line printer   A type of high-speed hard copy output device which prints a whole line at a time.

line segment   The portion of a line bounded by the endpoints in graphics.

line scan   The horizontal scan of a cathode ray tube picture or data display.

line tests   Photographic tests of pencil drawings for smoothness of animation.

line-up slides   Slides in register mounts specifically produced for lining up the projectors of a multi-projection system.

line-up switch   A switch on slide projectors or on a projector control unit, to turn on the lamps for lining-up purposes.

linker   In computer practice, a program used to combine separate *relocatable* modules produced by an assembler or compiler, to produce a single executable program for later use.

lining up   (1) Setting up a camera in rehearsal ready for actual shooting. (2) Accurately adjusting equipment for optimum performance. (3) Accurately setting up two or more slide projectors for multivision.

lip sync   Simultaneous picture and sound recording, to give exact correspondence between lip movement and spoken speech.

liquid crystal   A display device where reflection or transmission of external light is alterable by a control potential.

liquid gate   In motion picture printing, an aperture where the film is immersed at the time of exposure in a liquid of the same refractive index as the base, to avoid surface scratches on the negative being printed.

liquid printing   Also called *wet printing;* see *liquid gate*.

lith film   An extremely high contrast black and white film.

lith mask   (1) A mask produced by photographing artwork on *lith film* for *sandwiching* with a transparency in a slide mount. (2) A high contrast mask used for producing a composite image in optical printing on to a single slide.

live   Describing information processed as it occurs in *real-time*, as opposed to recorded.

live action   Shots of real motion as opposed to animation.

live mic   A microphone that is currently active.

lm   lumen, the unit of luminous flux.

load   (1) To apply an impedance across the output terminals of a source; hence, the impedance so placed. (2) In computer practice, to call in a program, usually from disk or tape.

lock   To arrange for two or more film or tape transports to be synchronised together.

lock-up   In a computer or micro-processor based device, to become useless by getting stuck in a state or a loop of states. The system may need to be reset (sometimes with information loss) to get out of this condition.

logarithm   Of a number, the power of a fixed number (called the 'base', usually 10), that equals the number.

logarithmic   A non-linear scale based on a logarithmic ratio; usually to the base 10 logarithm; see *bel* and *density*.

logger   Tape or cassette recorder with very low tape speed for continuous monitoring, e.g. the output of a broadcast station.

logic   (1) In computer practice, the use of program operators to perform logical operations on binary numbers, e.g. AND-ing them. (2) In electronics, an arrangement of *gates* and other circuit elements to perform a specified task.

long focus lens A camera lens having a narrower angle of view and greater magnification than is normal for the format.

long pulse A long cue-tone to indicate end-of-programme.

long shot Scene showing a general view of the setting within which the action takes place.

loop (1) A length of film or tape joined end to end to form a continuous band. (2) In a film mechanism such as a camera or projector, a length of free film between two driven points. (3) In computer terminology, a sequence of steps repeated a specified number of times. (4) In multivision control, a facility to repeat a section of the programme of cues as many times as may be specified.

LOR Laser Optical Reflection, the principle of a videodisc system.

LOT Laser Optical Transmission, the principle of a videodisc system.

loudspeaker A transducer for converting an electrical signal into sound.

low band A videotape recording system not meeting TV broadcast standards, as in the original *EIAJ* specifications for U-matic and other domestic systems.

low key The characteristic of a scene in which the majority of tones in the subject or its reproduction are at the darker end of the scale.

low-level language A language in which each instruction has a single corresponding *machine-code* equivalent.

low noise A relative term of no specific meaning, e.g. a low-noise transistor or low-noise magnetic tape; see *noise.*

low-pass filter Device to attenuate high frequencies.

LPC Linear Predictive Coding.

LRC Longitudinal Redundancy Check.

LTC Longitudinal Time Code, sometimes known as audio *time-code*, recorded in the audio track of videotapes in *off-line* editing.

LS Long Shot. A general view of a scene.

LS Loudspeaker.

LSI Large Scale Integration.

lumen Measure of light, as luminous flux.

luminaire Light fitting in studio or theatre giving some control of the colour, size, shape and sharpness of the light, e.g. *broad, flood* or *spot.*

luminance (1) The brightness of a surface emitting or reflecting light. (2) The video signal determining the brightness of the image, as distinct from *chrominance* which determines its colour.

luminous flux The rate of flow of visible light from a source; measured in *lumens*.

LUT Look Up Table, fixed data or addresses used within a computer program.

lux Unit of illumination, equivalent to the incidence of one *lumen* per square metre.

LVR Longitudinal (i.e. linear) Video Recording, in contrast to *helical-scan* recording.

lx lux.

M

m milli-, a prefix denoting a factor of one thousandth, 10^{-3}.

m metre

μ micro-, a prefix denoting a factor of a millionth, 10^{-6}.

M mega-, a prefix denoting a factor of one million, 10^{6}.

M II Trade name for VTR system of broadcast quality (Matsushita), using ½ inch metal tape for component recording of *luminance* and *chrominance* on separate tracks.

M and D (Motion Picture Film) *masters* and *dupes*.

M and E Music and Effects, a sound track containing music and effects without speech or dialogue.

M&S mic Middle & Side microphone, a combination used for stereo sound recording.

MAC Multiplexed Analogue Component coding of a colour television signal, using time division and compression; intended for *DBS* transmission and capable of extension to higher resolution.

machine code The lowest level of programming for a microprocessor. The computer instructions are in *binary, hexadecimal* or *octal*.

macro A collection of separate functions or objects grouped together under one identity. Thus a macro in a computer program brings into operation all the individual functions built into the macro, without having to mention them all again individually.

macro lens A lens capable of very close-up focusing.

magazine (1) A light-proof film container for use with a camera, printer or processing machine. (2) A compartmented slide carrier that holds slides ready for use in a projector.

magnetic cartridge A gramophone pick-up based on a moving magnet assembly.

magnetic film A strip of magnetically coated or striped material, having perforation holes similar to those of photographic film for transport and synchronisation.

magnetic flux The concentration of magnetic lines of force in a region, determined by the field intensity and the permeability of the medium; measured in Webers.

magnetometer Instrument for measuring magnetic fields.

magnetron Power valve used to generate microwaves.

mag-opt A motion picture print with both magnetic and photographic (optical) sound tracks.

main frame A full-size computer, as opposed to a mini- or microcomputer.

main title The section of a motion picture film in which the name of the production and the principal credits are displayed.

maltese cross Part of a mechanism providing intermittent frame-by-frame movement in a film projector. See *geneva movement*. See *Figure M.1*.

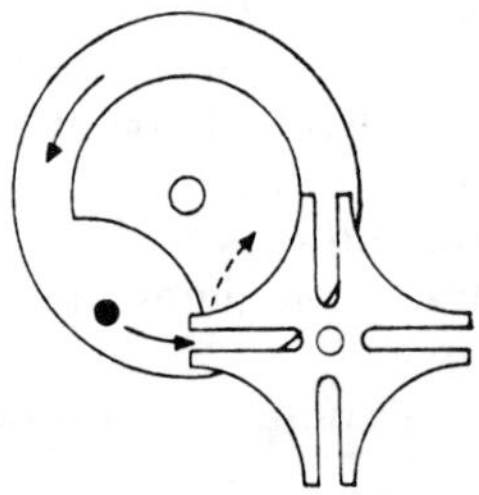

Figure M.1 Maltese cross mechanism

manchester code A commonly used data signalling code employed in magnetic recording and fibre optic communications.

Marata A rigid material for rear-projection screens, of fine grain and dark colour to minimise the effect of extraneous light. (Trade name).

marker In a graphics display, a user-defined symbol which can be invoked repeatedly.

mark it A command at live action filming to have the *slate* put in so as to identify the shot about to be taken.

married print A motion picture print with both picture and sound on the same strip of film correctly synchronised for projection.

marrying-up Preparing picture and sound negatives in the laboratory for making *married prints*.

masking (audio) Psycho-acoustic effect whereby a sound suppresses the subjective effect of noise in the adjacent frequency band.

masking (colour)   In colour photography, a method of correcting unwanted absorption in the dye components by the use of complementary colour images, which may be integral to the material or separate from it.

masking (contrast control)   Combination of a normal photographic record with a low density image of inverse tonal values to reduce its effective contrast in printing or projection.

masking (printing)   The indication of the unwanted areas of an illustration, either by means of an opaque cut-out overlay, or by lines marked on a transparency, or on the back of an illustration.

masking (projection)   Limitation of a projected picture by combination of the aperture in the projector and black borders surrounding the screen.

masking (video)   The ability to adjust the colour balance by matrixing the *RGB* signals. Used to match colour film primaries to the television standard.

master   (1) The original 16mm reversal film exposed in the camera, after processing. (2) A special positive print made from an original negative for protection or duplication rather than projection. (3) The final version of any programme (video, audio, or tape-slide) from which *show copies* will be made. (4) An output gain control on an audio mixer or video chain. (5) A similar control on a lighting desk.

match dissolve   A dissolve where one object is seen in different settings but occupying the same position on the screen throughout the dissolve.

matching transformer   A device for coupling two systems of differing impedance, e.g. matching a microphone to an amplifier input circuit.

match line   An animation term for a line showing exactly where an object on an upper *level* has to appear to pass behind one on a lower level.

matrix   An electronic circuit system used to combine several signal sources by different paths, particularly one providing a network of columns and rows.

matrix printer   A text and graphic printer using an electromagnetically operated matrix of needles, usually 7×9, producing characters or graphics as a pattern of dots.

matt   The property of a non-specular surface reflecting light uniformly in all directions.

matte   An opaque mask limiting the area of a picture which is exposed in special effects. It may be a cut-out aperture or a high density image on film, while in video it may be electronically generated to blank off the appropriate signal.

matte box   A front-of-lens device, usually rectangular in section, acting as a lens-shade and holding optical and other filters and shaped apertures for special effects.

MATV   Master Antenna Television. Term used for TV distribution from a central aerial in flats and apartment blocks.

maxwell   Unit of magnetic flux.

MCB   Miniature Circuit Breaker, a re-settable alternative to a fuse for subcircuit protection.

MCPS   Mechanical Copyright Protection Society (UK). The body which acts to protect the interest of writers/arrangers when works are mechanically reproduced.

MCR   Mobile Control Room.

MCS   Medium Close Shot, showing a subject about waist-length.

MCU   Medium Close-Up, showing head and shoulders of a subject.

MDS   (1) Microprocessor Development System. (2) Multipoint Distribution System in television.

MECL   Multi Emitter Coupled Logic.

media transfer   Copying from one medium to another, e.g. from tape-slide to video.

mega-   A prefix denoting a factor of one million, 10^6.

memory   Digital store for data, commonly consisting of integrated circuits.

menu   A list of options presented by a computer to its operator, usually on a visual display unit, as for example the list of picture control options available to an animator in a graphics package.

metal halide lamp   A compact mercury arc with metal halide additions, enclosed in a quartz envelope, often protected by a hard glass shield.

metallised screen   A screen whose surface has been treated with metallic particles giving substantially specular reflection.

metal oxide   Combination of a metal and oxygen.

metal tape   Magnetic recording tape with ultra-fine metal particles coated in a binder layer or vacuum deposited.

MF   Medium Frequency of radio waves, 300 kHz to 3 MHz.

M-format, M-wrap   The tape path used in the *VHS* system of *VCR*.

mho   Reciprocal of an ohm, a measure of conductivity.

MHz   Megahertz. One million cycles per second.

mic   Microphone.

micro   (1) A prefix denoting a factor of one millionth, 10^{-6}. (2) abbreviation for microprocessor or microcomputer (slang).

microbreak   A very short supply interruption.

micro cassette   A miniature cassette similar to a compact cassette but about one third the size, used for dictation, music or data recording.

microdrive A miniature data cassette containing endless tape. (Trade name).

microfilm Photographic reproduction of data and/or illustrations in a size too small to read without magnification.

micron Obsolescent name for one millionth of a metre, μm.

microphone A device to convert sound to an electrical analogue signal.

microprocessor (1) An integrated circuit chip containing a complete processor. (2) Loose name for a programmed micro-computer.

microsecond One millionth part of a second.

microwave Radio transmission above 1 GHz; used for transmitting TV pictures between different sites using dish aerials.

middle fairing A *fairing* used to blend animation movement from one speed to another.

mike Microphone.

milli- A prefix denoting a factor of one thousandth, 10^{-3}.

millisecond One thousandth part of a second.

mini floppy disk A 5¼ inch magnetic coated disk for data storage. (Full size floppy disks are 8 inch diameter.)

mips Million instructions per second.

mired MIcro-REciprocal-Degree. A unit of colour temperature, being one million divided by the *kelvin* value; thus daylight of 5500 K corresponds to 182 mired.

MIT Massachusetts Institute of Technology (USA).

mix (1) To blend audio or video sources together creatively. (2) A visual effect equivalent to a *dissolve*.

mix down The process of mixing audio signals from a large number of tracks to a smaller number, e.g. an 8-track master to a stereo pair of tracks.

mixing The process of combining separate sound or visual sources to make a smooth composite or change.

MLS Medium Long Shot, showing almost a full-length figure of a subject.

MMDS Multichannel Multipoint Distribution Service, a pay-TV system using microwave links rather than cable.

mnemonic code the elements of *assembler language*, short codes which mean more to the programmer than binary *machine code*. For example, DCR A, meaning Decrement the Accumulator, is used rather than 00111101 in binary or 3D in *hex*.

MNOS Metal-Nitride-Oxide-Semiconductor.

Möbius loop A length of tape or film joined to form a loop with a half-twist in it so that each surface will pass any point, used in some magnetic cartridges and printer ribbons; see *Figure M.2*.

MOD Minimum Operating Distance, in a camera lens specification.

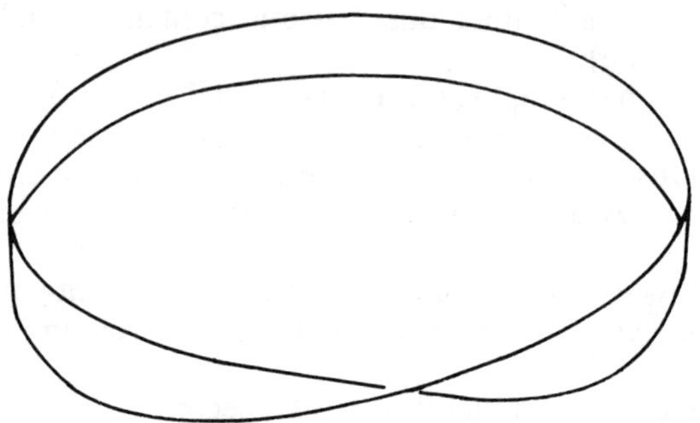

Figure M.2 Möbius loop

model sheet An animator's reference sheet showing specifications and relative sizes of cartoon characters.

model shot A film or video shot in which models are used instead of real objects.

modelling light Light(s) placed to reveal the form and texture of the object being photographed.

modem MOdulator/DEModulator; device used for sending digital information over audio circuits, such as telephone lines.

modified NTSC Colour television system using *NTSC* coding but with a colour sub-carrier frequency of 4.43 MHz (as used for *PAL*) instead of 3.8 MHz. This is sometimes more convenient for triple standard video cassette recorders and monitors.

modulation Alteration of amplitude, frequency or phase of an oscillation by a different frequency, e.g. to impress a signal on a carrier wave. Pulse width, pulse code and pulse position modulation are also employed.

modulation noise Noise which appears only when modulation is present.

modulation transfer function A measure of the performance of a lens or reproduction system based on its ability to represent detail at increasing frequencies.

modulator (1) A device for impressing a signal on a radio frequency carrier. (2) A device for producing audio effects where one sound is modulated by another.

module In audio-visual practice, a self-contained programme that may be used alone or be integrated as part of a larger programme.

moiré Visual patterns formed by interference between two sets of regular divisions, such as the combination of a TV raster with a striped object in the scene; can be caused by any beating between two frequencies; see *Figure M.3*.

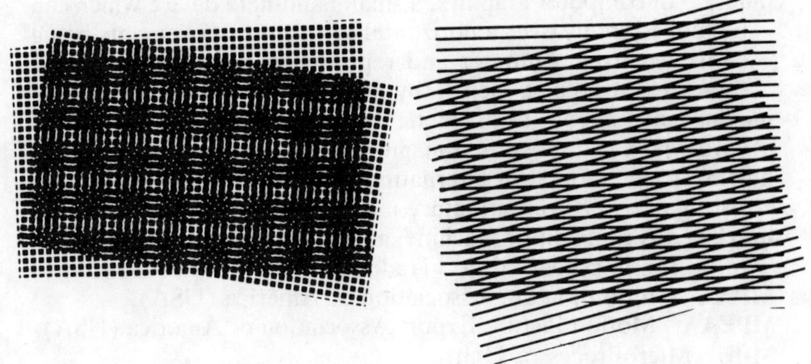

Figure M.3 Moiré patterns or lines formed by interference between two sets of regular divisions

monitor (1) A video display screen not fitted with radio-frequency receiving circuits, used to check the quality of what is being rehearsed or transmitted or to display text or graphical information. (2) A resident first-level computer operating system.

monitor loudspeaker A high-quality loudspeaker used to check results in sound mixing and recording.

mono Monaural, a single audio channel.

monochromatic light Light of a single wavelength (colour).

monochrome Reproduction in a single colour, normally as a black and white picture.

monophone A single earphone with a handle, often used at exhibitions.

monostable An electronic circuit element, with one stable and one unstable state. If triggered into the unstable state it will regain the stable state after a defined period. See *re-triggerable*.

montage A complex series of rapidly changing images and visual effects, often with several pictures seen simultaneously on the screen.

monotonic A signal changing only in one direction, not going back on itself in error.

MOS Metal-Oxide-Semiconductor.

mosaic The pattern of elements in a CCD image sensor.

MOSFET Metal-Oxide Field-Effect Transistor.

motorboating Low-frequency noise which may be caused by instability in a sound amplifier, or by the influence of perforation holes on the sound track of 35mm motion picture film.

mount A carrier for a still film frame or transparency. See *offset mount* and *register mount*.

mouse   In computer graphics, a small hand-held device which can be moved freely over a horizontal surface, its movements being conveyed to a computer and reproduced on the cathode ray tube. It is used as a 'pencil' by the animator.

moving coil   An electromagnetic transducer consisting of a coil of wire which can move in a magnetic field.

moving iron   A type of alternating current electrical meter consisting of a soft iron armature moving within a coil of wire.

moviola   A film editing machine, allowing picture and sound films to be run in synchronism. (Trade name).

MPAA   Motion Picture Association of America (USA).

MPEAA   Motion Picture Export Association of America (USA).

MPU   MicroProcessor Unit.

MPX   Multiplex.

MS   Medium Shot, generally showing a subject from the waist up.

MSB   Most Significant Bit, in a digital word.

MSI   Medium Scale Integration.

MSO   Multiple System Operator, an organisation with a number of *MATV* systems.

MTBF   Mean Time Between Failures. A reliability assessment.

MTF   *modulation transfer function.*

multicore   A cable with a number of separately insulated conductors within one sheath.

multi-frequency control   A control system where different channels are controlled simultaneously over one tape track by the use of several frequencies, sometimes coded in pairs.

multi-image   A presentation with several projectors on any screen area (USA term).

multimedia   A presentation employing a mixture of media, e.g. tape-slide, motion picture, video and live action.

multiplane animation   Cartoon animation with *cels* placed on several planes at different distances below the camera.

multiplex   A time-sharing process whereby several different signals are transmitted over a single signal path.

multiplexer   Device to enable images from several sources (slides or motion picture) to be transferred on to one medium such as video or film.

multi-program   Several programs running in a computer at the same time by time-sharing the processor.

multi-screen   An audio-visual system having a number of image areas presented simultaneously to the viewer.

multi-tasking   A computer system able to give the illusion of executing more than one task at a time.

multi-track   A magnetic recorder having four or more parallel tracks. Commonly 4, 8, 16 or 24 tracks.

multi-tracking   (1) A technique of sound recording with a separate track for each source to permit subsequent mixing and blending. (2) Building up an audio track by the addition of several successive stages.

multi-user   A computer system capable of interacting with more than one user at a time.

multivision   General term for multi-screen or multi-image presentation.

Mu-metal   Magnetic screening material. (Trade name).

Munsel system   A method of defining colours.

MUSE   Multiple Sub-Nyquist Sampling Encoding, Japanese bandwidth compression system to accommodate *HDTV* transmission within an existing satellite channel.

music chart   A breakdown, frame by frame, of a music track so that an animator can work exactly to the beat.

music-track   An audio track that carries only music.

mute   (1) A picture record only, without associated sound track on the same film or tape. (2) To disconnect or suppress the sound.

Mylar   Trade name for polyethelene terephthalate, a stable material sometimes used as a base for magnetic tape.

N

n   nano-, a prefix denoting a factor of 10^{-9}.

N   Newton, the unit of force.

NAB   National Association of Broadcasters (USA). Term used for standards specified by that organisation.

NAB cartridge   Broadcast standard endless loop magnetic tape cartridge standardised by the *NAB*, made in three tape capacity sizes.

NAB spool   Magnetic tape spool with large central hole standardised by the *NAB*, available in a range of outer diameters.

NABT   North American Broadcast Teletext system (US).

Nagra   Trade name for very portable magnetic tape recorders, often with film camera sync capability.

NAND gate   A gate which gives an output equal to 0 when all its inputs are 1. (*Positive logic* convention).

nano-   A prefix denoting a factor of 10^{-9}.

nanosecond   One thousandth part of a microsecond.

narration   see *voice over*.

NARTB   National Association of Radio and Television Broadcasters (USA).

NATO National Association of Theatre Owners (USA).

NATTKE National Association of Theatrical, Television and Kine Employees (UK).

NBC National Broadcasting Company (USA).

NBS National Bureau of Standards (USA).

NC Normally Closed, a relay or switch contact closed in the un-operated condition.

NCTA National Cable Television Association (USA).

ND Neutral Density filter, used to reduce light level without affecting colour.

negative An image or condition in which the tones or impulses are the reverse of the original.

negative logic Logic convention with a 'one' being represented by a negative potential or no potential.

negative resistance A resistance condition where the current decreases with increasing voltage across the resistor (or circuit).

neg-pos Photographic system in which the film exposed in the camera is processed as a negative image and used to print a separate positive.

neons A term in rostrum camera work for neon-like lettering effects.

nested structures A computer programming technique, where one *loop* or decision structure is placed within another, so that the inner one is *executed* as often as required by the outer one.

net A logical linking of fixed points (*pins*) in a computer-aided design display.

neutral density Having substantially equal values of density to the whole spectrum of visible light.

Newvicon TV camera tube, more sensitive than Vidicon but with reduced lag. (Trade name).

Newton's rings Optical interference patterns caused when two surfaces separated by a very small distance are illuminated, for example, the film surface of a slide and its cover glass.

NG No Good, a common notation for a filmed take or recording, to indicate it is not to be printed.

NHK Nippon Hoso Kyokai (Japan Broadcasting Corporation).

Ni-Cad Nickel Cadmium rechargable cell or battery of cells. (Trade name).

nit An obsolete unit of light measurement, equivalent to one *candela* per square metre.

NLQ Near Letterpress Quality, describing performance of alpha-numeric print-out systems.

NMOS N-channel Metal-Oxide-Semiconductor.

NMR Nuclear Magnetic Resonance.

NO Normally Open, a relay or switch contact open in the un-operated condition.

No-break Power supply system which will maintain power without interruption in the event of supply failure.

nodal head Mount for a motion picture camera providing rotation and tilt centred exactly on the *nodal point* of the objective lens.

nodal point The effective optical centre of an objective lens through which all rays may be regarded as passing.

noddies In an interview, reverse shots of the interviewer used as *cutaways* to bridge jump cuts.

node A point at rest in a vibrating system; a central point in a system.

noise Unwanted sound or signals in a communication system, especially random electrical energy interfering with audio and video.

noise filter Commonly a *low-pass* filter for suppressing noise outside the desired audio spectrum, but may be *bandpass* or *high pass*.

noise floor Inherent noise level.

noise gate Automatic switch which disconnects an audio-frequency path when the signal falls below a preset threshold level.

noise reduction Circuit or system for reducing subjective effect of noise on sound or picture quality.

noise slice Method of reducing *noise* in a TV signal by removing the information in a TV waveform between predetermined levels, e.g. below 5% of signal amplitude (*black slice*); also termed *noise coring*.

non-composite video signal A video signal that contains picture and blanking information but not synchronising signals.

non-equivalent gate See *EXCLUSIVE-OR gate*.

non-flam Alternative term for Safety Film.

non-volatile store A data store which retains data when power is off.

NOR gate A gate which will produce an output state 0, when any of its inputs are at a 1 state. (*Positive logic* convention).

north/south/east/west convention A convention referring to the top of the field of an animation scene as North, the right hand edge as East, etc.

NOS Nederlandische Omroep Stichting, Netherlands state broadcasting organisation.

notch A shallow cut-out made on the edge of a strip of motion picture film to actuate the light change in a printer, or to stop the film in certain forms of automatic presentation.

notch filter A filter which rejects a single frequency, commonly used in measurement of total harmonic distortion.

NPL National Physical Laboratory (UK).

NPN   A form of transistor.

NRK   The Norwegian state broadcasting organisation.

NRZ   Non-Return to Zero. (cf. *RTZ*).

NTC   Negative Temperature Coefficient, e.g. resistance falls with rising temperature.

NTP   Normal Temperature and Pressure. A standardised condition of zero degrees Celsius and 760 mm of mercury.

NTSC   National Television Standards Committee, hence the American colour TV coding system, 525 lines, 60 fields, 30 frames per second.

null   (1) To adjust to the lowest value or the exact centre value, e.g. setting an *op-amp* for least output with zero input. (2) A *string* space reserved in memory but containing nothing, i.e. of zero length.

number board   See *slate*.

O

OB   Outside Broadcast.

objective   The main lens system of an instrument, such as a camera, producing an image of the object presented.

OCR   Optical Character Recognition, capture of characters as data rather than graphics, to permit editing and further word processing.

octal   (1) Numeric representation to the base 8, rather than decimal base 10. (2) An eight-pin circular polarised connector used for valves.

octopus cable   A cable with multiple plugs.

octave   An interval representing a doubling of frequency.

Oe   *oersted*.

OEM   Original Equipment Manufacturer.

oersted   Unit of magnetic intensity.

off-air   Reception of a broadcast signal.

off-line   Isolated, not connected to the line, central computer system, etc.

off-line edit   Editing video material using low-cost equipment to produce a *rough cut* prior to using expensive broadcast standard equipment for the final work.

off-scale   (1) Beyond the range of an instrument. (2) In motion picture printing, outside the range of the normal light point scale of a printer.

offset   A positive or negative time displacement in systems employing time code synchronisation.

offset lenses Slide projector lenses in mounts designed to allow the lens axis to be out of line with centre of the slide; for keystone-free projection when two or more projectors cover the same screen area.

offset mounts Slide mounts in which the aperture is not in the centre of the mount, used to avoid keystone distortion when two or three projectors share a common screen area.

ohm Unit of electrical resistance.

OHP Overhead Projector.

OIRT International Radio and Television Organisation.

omega wrap Tape path on the drum of a helical-scan videotape recorder giving slightly less than 360° of contact; see *Figure O.1*.

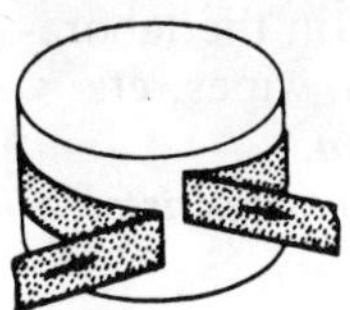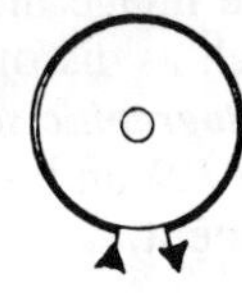

Figure O.1 Omega wrap. The tape in the helical-scan path makes almost a complete circuit of the recording head drum

omnidirectional Equally sensitive in all directions.

OmniMax Development of the *IMAX* system using the same size of film but presenting an even wider angle of view on a spherically curved screen.

on-line 'Live', actively linked to the line, system, etc.

one light A motion picture print made at a single level of exposure, without alteration from scene to scene.

one-time rights Permission to reproduce one illustration in one edition of one publication in one language, in a specified, agreed-to area.

one-to-one position Camera setting where the object and the image are the same size.

OOO Out Of Order.

OOV Out Of Vision.

op-amp Operational Amplifier.

opaquing Painting cels after they have been traced.

opcode The mnemonic operation code for a microprocessor which specifies the elemental operation to be performed.

open loop control A control system not self-corrected by feedback.

open loop gain The *gain* of a closed loop system measured with the feedback path opened.

open reel A tape transport system with separate feed and take-up spools, as distinct from the enclosed path of cassette or cartridge.

operand That part of a microcomputer instruction which identifies the data to be operated on by the *opcode*.

operating level An audio signal level at +4 dBm equivalent to zero VU; see *VU meter*.

operating system A fixed program which is used when operating a computer.

ops Operations per second.

optical axis The axis of symmetry of the components of an optical system.

optical printing A method of photographic printing in which an image of the original is formed on the print stock by a copy lens.

optical ROM Laser disc technology for very high density storage of read-only data.

opticals Modifications to the picture image made in the laboratory after filming is completed, such as dissolves, wipes, etc.

optical sound Preferred term is *photographic sound*.

opto-isolator A combination of an *LED* and a *phototransistor* or photo*darlington* in a sealed component.

Oracle *ITV* teletext system (UK).

ordered dither In a graphic display, setting the intensity level (or colour) for each *pixel*, determined by a set of threshold values applied to the pixel array.

OR gate A gate where an output of 1 is produced if any input is 1. (*positive logic* convention).

ORF Österreichischer Rundfunk, Austrian state broadcasting organisation.

orientation (1) The alignment of the long axis of magnetic particles in magnetic tape manufacture to obtain maximum sensitivity; the direction of alignment differs according to the tape usage for audio or video recording. (2) The film position in 16 mm projection, emulsion-to-light (Type A) or emulsion-to-lens (type B). *Figure O.2*

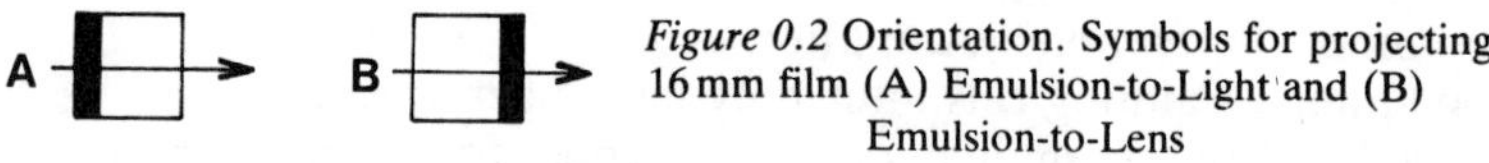

Figure 0.2 Orientation. Symbols for projecting 16 mm film (A) Emulsion-to-Light and (B) Emulsion-to-Lens

original (1) The film exposed in the camera, after processing. (2) The first video recording prior to copying or editing. Also a final edited master.

ortho Orthochromatic, sensitive to only the blue and green regions of the visible spectrum.

oscilloscope A cathode ray tube device to display and measure signals; it can give a graphical display of voltage against time.

OTF Optical Transfer Function. A method of accurately measuring the characteristics of lenses or photographic reproduction systems.

OTS Orbital Test Satellite.

out-of-sync When picture and sound do not synchronise.

out-takes Alternative takes that are not used in the final editing.

overcranking Filming at a higher speed than the intended projection speed, to slow down the action.

overdamped A damping system in which a steady state is reached in less than one oscillatory cycle; see *Figure D.1*.

overdub The addition of another musical part to an already recorded performance.

overhead projector A presentation device with a large horizontal projection aperture, with a lens and mirror/prism above. The image is projected above and behind the user's head.

overlap Extending the sound track into the next scene for smooth continuity.

overlay (1) The superimposition of one image on another without the background showing through. (2) A foreground piece in animation. It is registered on an upper *level* so permitting objects registered on a lower level to pass behind it without resorting to *match lines*. (3) In computer practice, where a program too long for available memory is processed by instalments, each segment 'overlaying' or replacing part of the previous code.

overload (1) To apply too large a signal to a system. (2) To draw too much power from a supply.

overmodulate The input of a signal amplitude greater than the normal operating range of the system can accept; see *Figure O.3*.

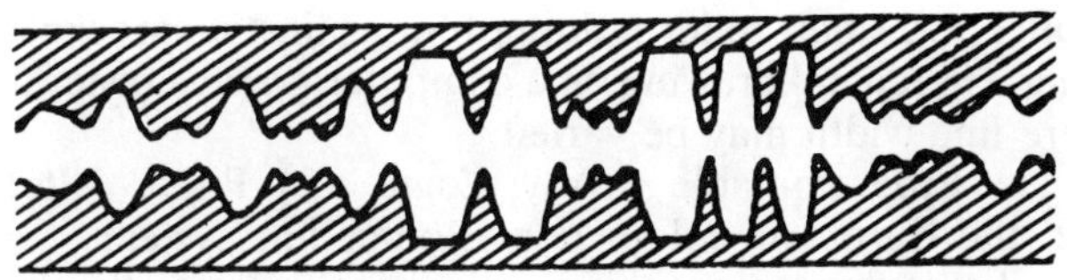

Figure O.3 Overmodulate. The amplitude of the peaks of the recorded signal are greater than the modulated track width available

over-range A situation where the input to an instrument exceeds its display capacity.

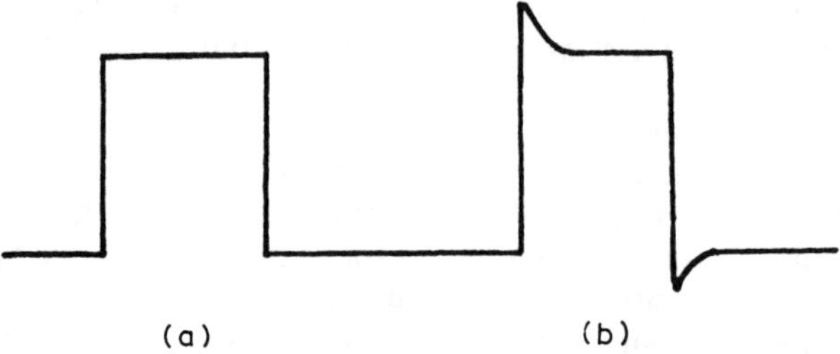

(a)   (b)

Figure O.4 Overshoot. The input pulse (*a*) is distorted at its
leading and trailing edges (*b*)

overshoot   A transient excess response at the beginning and end
of a pulse signal. In a video display it appears as a light or dark
line at the edge of an image where a marked change of
brightness occurs. (cf. *ringing*). See *Figure O.4.*
oxide   Ferric oxide particles coated in a binder on a base for
magnetic tape or disc.

P

p   pico-, a prefix denoting a factor of 10^{-12}.
Pa   Pascal, a unit of pressure.
PA   Public Address, sound reinforcement system.
package   A set of programs for a specific task on a computer. The
package may include special *hardware* as well as *software.*
pad   (1) A fixed value attenuator (2) In computer graphics, a *data
tablet.*
pad roller   A small roller holding film against a sprocket.
painting   Freehand drawing on a computer graphic raster display,
where line width may be varied.
PAL   (1) Programmable Array Logic. (2) Phase Alternation
Line, the European colour television coding system.
palette   In computer graphics and vision mixers, the range of
pre-determined colours from which a selection may be made.
pan   (1) To rotate a camera in a horizontal direction. (2) In
animation, movement of the field in any direction. (3) See *pan
pot.*
pan and tilt head   Mounting for a film or TV camera on its tripod
or pedestal which allows rotation in both horizontal and vertical
planes.

Panaglide Trade name for a harness with stabilised camera mounting giving smooth movement of a hand-held camera.

pancake (1) A reel of tape on a *core*. (2) A low platform used to raise artistes or objects a few inches off the studio floor to improve composition.

panchromatic Having a response to all colours of the visible spectrum.

panorama In slide projection, a single complete picture covering the whole or several parts of a multi-screen presentation.

pan pot Potentiometer control for moving ('panning') and locating the apparent source of sound within the area of stereo sound presentation.

parabolic reflector A concave reflector whose cross-section along the axis is a parabola, used to focus light, sound or RF radiation.

parallax The apparent difference in object positions resulting from a change in the point of observation. In camera work, the difference between the composition of the objects in the picture as seen through the viewfinder and through the camera lens.

parallel communication A multi-wire or *bus* system of communication.

parameter A number or value used in a specification or passed to a procedure.

parametric equaliser Equaliser that allows control of frequency and bandwidth.

parity In digital practice, the addition of a check *bit* to enable errors to be detected at *byte* level. A simple parity bit indicates if the sum of the number of ones in the preceding byte is odd or even.

PASCAL A high level language which when compiled produces *P-code*, a pseudo *machine code* which can be run on any computer with a P-machine interpreter.

paste up An assembly of different visual or text components on to a common base to form a coherent whole.

patch (1) A transparent piece of thin film used to repair a break in a film. (2) A small program inserted to remedy a defect in a computer program.

patch panel A *jackfield* to allow different outputs to be connected to different inputs.

pause Temporary programme stop; in particular, interruption of a tape transport mechanism without changing the function modes such as record or play back.

pause control Control on a tape recorder which temporarily stops the tape transport without unthreading, but does not necessarily give a still picture on a videotape recorder.

pay TV Non-broadcast television distribution system, generally using a cable network, where a charge is made for the programme selected for viewing.

PBS Public Broadcasting System (USA).

PC Printed Circuit.

PCB Printed Circuit Board.

PCM Pulse Code Modulation.

P-code Intermediate machine-independent code, produced by a PASCAL *compiler.*

PD Potential Difference (between two points in a circuit).

PDA (1) Pulse Distribution Amplifier. (2) Post Deflection Acceleration in cathode ray tubes.

peak programme meter A maximum signal level meter—usually in audio level monitoring.

peak-to-peak The amplitude of a waveform measured from positive peak to negative peak.

peak white Maximum amplitude of a video signal corresponding to the brightest white area possible on the screen.

pearlescent Cinema screen material with slightly translucent coating.

pedestal (1) A camera support with adjustable height. (2) In a television signal, an artificial black level inserted in a circuit to raise the black level of the picture a very small amount above the blanking level; see *Figure P.1.*

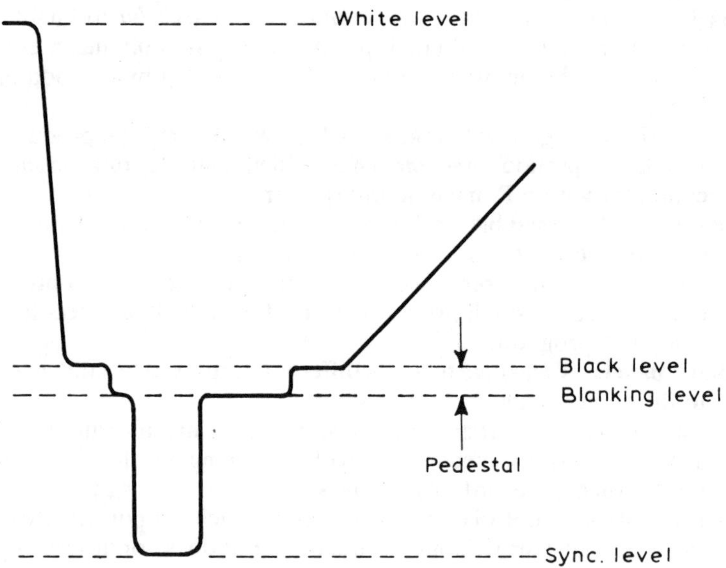

Figure P.1 Pedestal. As shown in a single line TV waveform

ped up, ped down   Movement of a camera vertically up or down.

peek   A *BASIC* instruction to a microcomputer to read a specified memory location.

peg animation   Animation in which the illusion of motion is provided by substituting artwork every frame or every second frame. The artwork is registed by pegs.

peg bar   Metal pins allowing artwork and captions to be accurately registered.

pencil test   See *line test*.

Perceval   Belgian teletext system.

perforated screen   A projection screen with small holes to permit a loudspeaker system to operate efficiently from the rear side.

perforations   Holes along the edge of a strip of film used for its transport and registration.

Performing Rights Society Ltd   An association of composers, authors and publishers for the collection of copyright fees (UK).

peripherals   External items added to a computer system, e.g. floppy disk drives, printers, visual display units, etc.

Perlux   A projection screen material of high reflectivity over a wide angle (Trade name).

persistence   Time taken for an image (usually on a cathode ray tube screen) to die away.

perspective control lens   An objective designed to be moved off-axis so that more than one projector can produce an image free from keystone distortion at the same part of the screen. See *Figure P.2.*

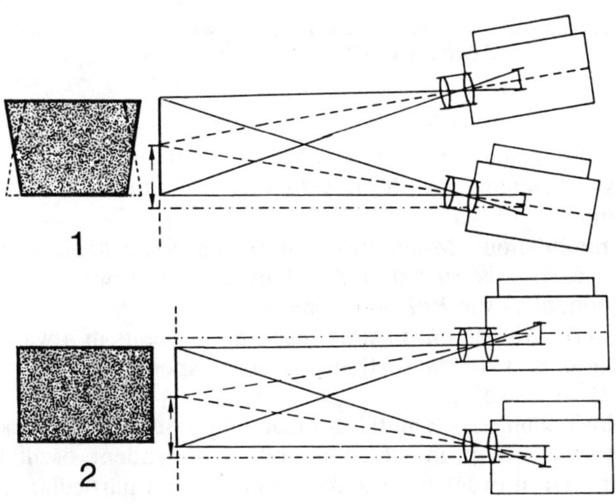

Figure P.2 Perspective control lens. Keystone distortion introduced by tilted projectors (1) is avoided by movement of the lenses off-axis (2).

PFL *pre-fade listen.*

phantoming A technique for sending more than one signal per pair of wires in a cable. See *phantom power.*

phantom power Power fed to a microphone using the audio signal cable without interfering with the audio signal. Also used in TV camera cables.

phase (1) The relation between two waveforms of the same frequency; phase displacement may be expressed by angular difference. When the cycles coincide exactly they are termed 'in phase'. (2) The angle between two vectors; see *Figure P.3.*

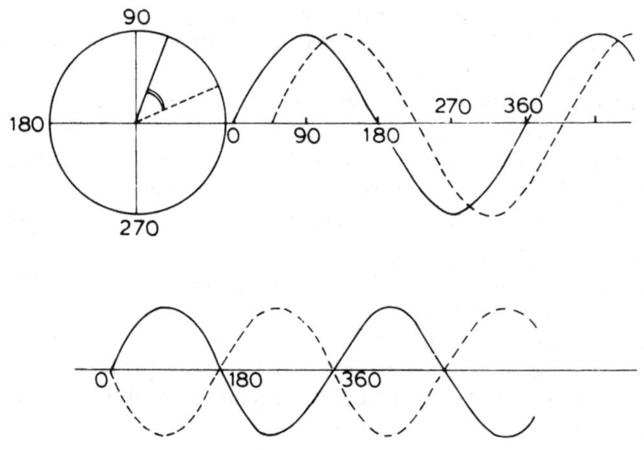

Figure P.3 Phase. Angular difference between two waves of the same frequency; a difference of 180° is termed 'anti-phase'.

phase control Control of phase angle. A method of altering the moment of switching a *thyristor* or triac, relative to the phase of the supply mains in order to provide power control, as for lamp dimming.

phase modulation Modulation where the phase angle of a sine wave carrier is varied, e.g. as used to transmit the colour component in the *PAL* video signal.

phase shift Movement in time resulting in points in a waveform being earlier or later than the corresponding points of a reference waveform.

phase-lock loop Automatic control of phase by feedback. An electronic circuit in which a voltage-dependent oscillator is controlled in order to lock the oscillator to a particular signal.

phasing (1) Audio special effect created by recombining a split signal after delaying one part by a very small amount. Varying the delay alters the frequency-dependent phase cancellations to create the effect. (2) Checking that loudspeakers are in phase with one another (rather than in anti-phase).

phon   Unit of loudness, on a scale whose zero is the threshold of hearing.

phonoplug   A small coaxial single-pin audio connector, also called an RCA connector. (Trade name).

phosphor   Substance emitting visible light of specific wavelengths when irradiated, as for example by an electron beam in a cathode ray tube or by ultra-violet radiation in a fluorescent lamp.

phosphor decay   The time taken for an image to die away after the electron beam ceases to write.

photodiode   Semiconductor device used to convert light into an electrical signal.

photoelectric cell   Device for converting light into an electrical signal.

photographic sound   System of recording and reproducing on motion picture film in which the sound track takes the form of variations in the density or width of a photographic image.

photometer   Instrument used for measuring luminous intensity of lights.

photomultiplier   A very sensitive type of vacuum tube photoelectric cell with built-in high gain amplification based on secondary emission.

photosensitive   Sensitive to light.

phototransistor   Transistor able to convert light into an electrical signal with amplification.

pick   The selection of a display item in computer graphics.

pick-up   A transducer for reproducing recordings on audio and video discs.

pick-up arm   A device for supporting a *pick-up*.

pick-up tube   A video camera tube.

pico-   A prefix denoting a factor of 10^{-12}.

picosecond 10^{-12} second. One-thousandth part of a *nanosecond*.

picture search   The ability to locate a frame on film, videotape or disc over a range of speeds.

pie chart   A circular form of graphical data representation; see *Figure P.4*.

piezo-electric   The interchange of electric potential and mechanical deformation.

pilot pin   A pin in a motion picture or rostrum camera or printer mechanism, engaging a perforation hole for precise registration during exposure. See *register pins*.

pilot tone   Speed control/camera sync system. The tone may be

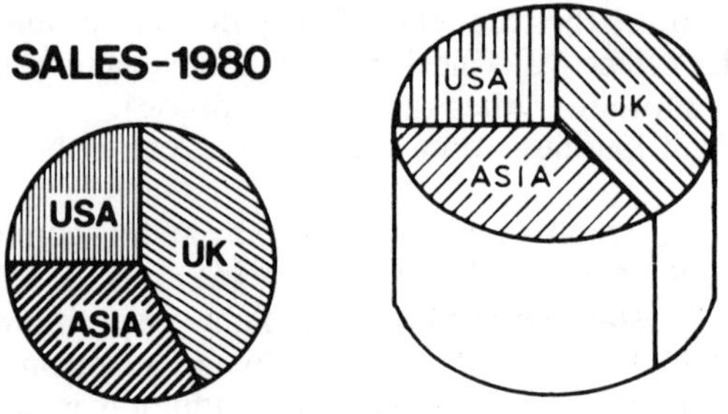

Figure P.4 Pie chart.

recorded on a full width tape using two narrow tracks in anti-phase.

pin   (1) The component of a camera or projector mechanism which engages in the perforation hole to locate the frame. (2) In computer graphics, the connection points on logic elements and components in *computer aided design* displays.

pin registration   Exact location of a film frame or image, as in a camera gate or slide mount, by means of pins engaging the perforation holes.

pin-cushion distortion   Image distortion in an optical or video system causing a rectangle to appear with concave sides and elongated corners. (cf. *barrel distortion*); see *Figure P.5*.

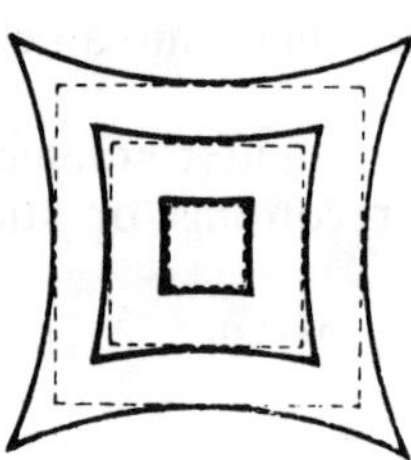

Figure P.5 Pincushion distortion.

pinhole   A small defective spot in a photographic emulsion or image, or in a magnetic coating.

pinhole slide   A slide-mount with a small hole in the middle, used in projector lamp and mirror adjustment by forming an image of the filament and its reflection on a translucent lens cap.

pink noise   *white noise* filtered at 3 dB/octave to provide constant energy per octave bandwidth.

PIO   Programmable Input/Output device.

pipping   The copying of data from one computer disk to another.

pitch (1) Distance between successive regularly occurring points, such as film perforations or video tracks. (2) The frequency of a sound, usually of one within the audible range.

pitch changer A device to alter the pitch or frequency of an audio signal. Used with only a small change of pitch, to reduce the possibility of *threshold howl*.

pitch control A continuously variable speed control on magnetic or disc audio reproducers, allowing the replay to be tuned exactly to musical instruments.

pixel The smallest element of a raster display; a picture cell with specified colour and/or intensity level.

pixillation (1) Visual effect in film and video where moving action is represented as a series of stills each held long enough to be recognised as static. (2) Also in video, an effect in which the whole picture is reproduced as a comparatively small number of enlarged *pixels*. (See *tile*).

PLA Programmable Logic Array.

plasma panel A display system, comprising a matrix of gas-filled cells which can be turned on and off individually.

plate A photographic print, motion picture film or still transparency, used as the background scene in special effects such as *back projection, reflex projection, aerial image photography* and *travelling matte* shots.

platen A surface providing support or contact, such as (1) The glass pressure plate for *cels* and background on an *animation stand*. (2) The working surface of an *overhead projector*. (3) The paper supporting surface on a computer printer.

platter Large horizontal disc used to support long rolls of film in continuous projection systems.

playback Reproduction of a recording.

PLL *phase lock loop*.

PL/M Programming Language for Microcomputers.

plotter A device for plotting graphical hard copy from a computer; see also *X-Y plotter*.

PLUGE Picture Line-Up Generating Equipment. A signal used to set gain and lift on monitors.

Plumbicon A lead oxide vidicon tube. (Trade name).

PMOS P-channel Metal-Oxide-Semiconductor.

PMT Photo Mechanical Transfer. A term used for the reproduction of existing material.

PNP A form of transistor.

podium A dais or raised platform used by a presenter or lecturer.

point (1) The exposure increment used in a motion picture printing machine, generally corresponding to a change of 0.025 log E. A printer point scale of 1 to 50 is normally employed. (2)

In typography, the height of characters expressed in units of 1/72 of an inch.

PO jack   A two or three circuit jack of ¼ inch, 6.25 mm nominal diameter.

poke   A *BASIC* instruction to a microcomputer, to set a specified memory location to a specified pattern of states.

Polacoat   A material for fine-grain rear-projection screens, available in glass, rigid plastic, or flexible plastic. (Trade name).

pola filters   Small sheets of *Polaroid*, sometimes mounted in glass, used for fitting in front of the camera lens.

polar diagram   A plan view of the response pattern of a microphone, loudspeaker, or aerial system. Also applicable to light sources and projection screen reflection or transmission characteristics; see *Figure P.6*.

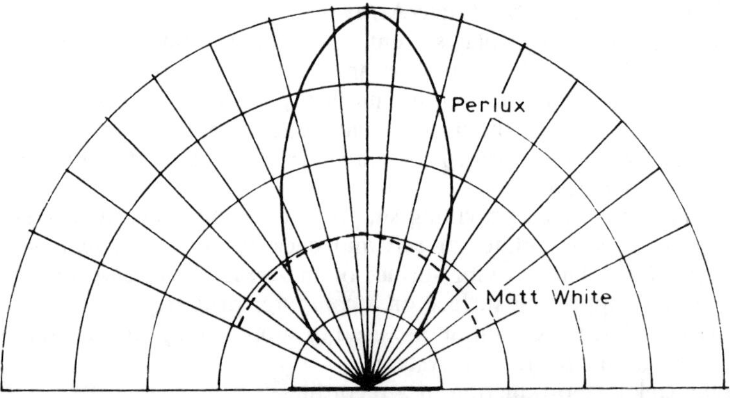

Figure P.6 Polar diagram. An example showing relative directional reflectivities of matt white and Perlux screens.

polarisation   Modification of electromagnetic radiation, especially light, so that its vibration takes place in a plane or a limited series of planes.

polarity   Of positive or negative electrical state.

Polaroid   (1) A transparent plastic material capable of polarising visible light (Trade name). (2) An instant-picture photographic system. (Trade name).

Polascreen   A transparent, neutral coloured filter which polarises transmitted light. (Trade name).

polecat   Spring-loaded pole which can be mounted between floor and ceiling for rigging on location.

pole tips Those parts of a magnetic head which protrude beyond the circumference of the head drum to make contact with the tape.

port A multi-wire input/output to a computer. Usually as many wires wide as the computer word, typically 8.

portapack Portable battery-operated video camera and recorder.

portrait Picture image format with the major dimension vertical.

positive (1) Polarity of electrical signal (+). (2) Photographic image in which the tones (and colours where they appear) reproduce those of the original scene.

positive logic A logical system in which a 1 or ON state equals a positive voltage, commonly 5 V, and an 0 or OFF state equals no voltage.

post production Editing of pre-recorded material, including use of special effects and audio dubbing.

post sync Post-synchronising, recording synchronous dialogue or sound effects for a scene after it has been photographed.

posterisation Picture reproduction using only a few specific tones or flat colours, with most of the image gradation and detail suppressed.

potentiometer A variable resistor where the wiper and the two ends are employed as a potential divider.

POV shot Point-Of-View shot, as though seen by the actor.

power amplifier An electronic amplifier which accepts a voltage input and provides a high power output, commonly used for driving loudspeakers.

power factor The phase relationship of voltage and current in an AC circuit. Unity power factor represents a non-reactive load.

p-p *peak-to-peak*.

ppm parts per million.

PPM (1) Peak Programme Meter used to monitor audio levels. (2) Pulse Position Modulation.

pps Pictures per second, film or video frame speed.

PPT Projected Picture Trust (UK).

preamble/postamble In digital recording, sync and identification groups before and after the main information.

pre-amplifier First stage of amplification.

pre-echo Unwanted hearing of signal before it should have arrived; can arise from adjacent grooves of disc records, or from magnetic transfer between adjacent turns of wound magnetic tape.

pre-emphasis Control of frequency response before transmission or recording.

pre-fade listen Facility on a sound desk allowing channels to be heard before fading up.

première   The first major public presentation of a motion picture, video or multiscreen production.

pre-recorded   Material for inclusion in a programme that is already in a recorded form.

pre-roll   Time required after starting a projector, telecine or videotape recorder to provide stable pictures and sound.

presence   Boosting the frequency response of an audio circuit, between 3 kHz and 8 kHz, to create an illusion of nearness to the sound source.

Presfax   BBC communication service using *ICE* to provide regions with programme scheduling information.

pressure pad   A component which ensures good contact between the surface of magnetic tape and the face of the head.

pressure plate   That part of a motion picture camera, printer or projector which holds the film flat at the time of exposure or projection.

Prestel   British Telecom viewdata service. (Trade name).

preview   (1) In video, the facility to see a picture before transmission or recording. (2) Special presentation of a completed motion picture production prior to its exhibition to the general public.

PRF   Pulse Repetition Frequency.

primary battery   A primary cell converts chemical energy into electrical energy; the process is not reversible and the cells cannot be fully recharged. A primary battery is an assembly of such cells.

primary colours   The essential components in a colour reproduction system from which all available hues can be produced by mixing. In a three-colour additive process, such as a colour TV tube, they are red, green and blue. In a subtractive process, such as colour film, they are yellow, cyan (blue-green) and magenta (red-purple).

primary (transformer)   The input winding of a transformer.

prime lens   The *objective* of a camera or projector, normally of fixed focal length, to which a supplementary attachment or range extender is added.

print   Photographic copy of a film, usually with a positive image.

printer   (1) Computer peripheral device to produce hard copy, i.e. on paper. Usually alphanumeric, but some printers can handle graphics. (2) Machine for the exposure of film to produce photographic copies.

printer lights   Figures representing the levels of exposure required in a motion picture printer, usually quoted for red, green and blue on a scale of 1–50.

printout   The printed output of a computer. Also called *hard copy.*

print through   Unwanted transfer of the signal on a magnetic tape record to adjacent layers of the tape with which it is in contact when wound on a spool.

prism   Optical component in the form of a transparent solid with inclined plane surfaces which re-directs a light beam by refraction and/or internal reflection.

probe   A tool used to gain access to signals in an electronic unit under test, with as little disturbance as possible.

proc. amp   Processing amplifier used to 'clean up' a video signal.

process shot   (1) General term in motion picture production for a trick shot created by special effects photography and/or optical printing. (2) A studio shot in which the background is a still or moving projected picture.

processing   The chemical treatment of exposed photographic material so as to render the latent image permanently visible.

production assistant   General assistant to the Director or Producer, responsible for scripts, *continuity* and logging of shots.

production master   Modified copy for production use.

program   The sequence of instructions called software that executes in a computer or microcomputer when it is run.

programme   The show material to be seen and heard when the production is presented.

programmed instruction   A method of instruction in which the student learns by following a controlled programme of instruction.

programming   Telling the computer what to do and in what order to do it, creating software.

projection television   Video presentation system in which the picture is optically displayed on a separate large screen, rather than on the face of a cathode ray tube.

projector   Device to project images on to a screen, from a motion picture film, transparent photographic slides or strips, or electronically generated video sources.

projectors   In computer graphics, lines passing through an object to intersect with the view plane of projection.

projector stack   Stand holding several automatic slide projectors, usually one above the other. Double and Triple Stackers are in use.

PROM   Programmable Read-Only Memory.

prompt   The character or string of characters on a *visual display unit* which lets the user know that input is expected.

prompter   Device used to provide script for on-camera presenters.

proportional spacing   In graphics, where the letters forming the text on a VDU screen, or being printed, are not of uniform width.

props Scenery property such as furniture, food, guns, newspapers, etc.

protocol A procedure agreed upon by agencies wishing to communicate.

PRS Performing Rights Society Ltd (UK), an association of composers, authors and publishers of musical works to collect public performance and broadcast royalties on behalf of its members.

PSC Portable Single Camera in video production, often a *camcorder*.

psi pounds per square inch.

psophometer A noise measuring meter.

PTC Positive Temperature Coefficient, e.g. resistance rising with rising temperature.

PTH Plated Through Holes (in a printed circuit board).

PTT Post, Telegraph and Telephone, usually a state communications department.

PU Pick-Up.

public address An audio system for presenting speech or music to a large audience.

puck In computer graphics, a hand-held input device used on a *data tablet* to enter coordinates by the use of programmable buttons.

pull back The backward movement of a camera away from its subject.

pull-down The operation of moving film from one frame to the next in camera or projector mechanism.

pull up A component in an electronic circuit to ensure a 'high' state in the absence of any active device dictating a 'low' condition. Usually a resistor, but can be an active or constant current device.

pulse Rapid change of state between two electrical levels and return, usually of short duration. It can also include a short burst of tone, but an A-V cueing signal is properly known as a *cue tone*.

pulse & bar Video test waveform.

pulse cross Technique for allowing examination of sync pulses on TV monitor.

pulse generator Device to generate a train of electrical pulses of pre-determined duration and frequency.

pulse-width modulation Signalling system where information is carried by means of coding wide and narrow pulses.

pup A small spotlight.

push-on, push-off Video transition effect in which the first scene appears to be pushed horizontally off the screen by the second scene. (Cf. *conceal/reveal, scroll*).

PVC   Polyvinyl Chloride.
PWM   Pulse-Width Modulation.

Q

***Q, Q*-factor**   A measure of the efficiency of a resonator or resonant circuit. High Q values represent a narrow width of resonance.

QA   Quality Assurance.

QIL   Quad-In-Line, packaging of an integral circuit with the mounting and contact pins in four staggered and parallel rows for flat mounting.

QPSK   Quadrature Phase-Shift Keying.

Q-signal   In the *NTSC* colour system, represents the ***chrominance*** on the green-magenta axis.

quadraphony   Four channel sound system.

quadruplex   A method of videotape recording using two inch wide tape with four record/replay heads mounted on a rotating headwheel, which traverses the tape at an angle slightly greater than 90° to the edge of the tape. (Trade name.) See *Figure Q.1*.

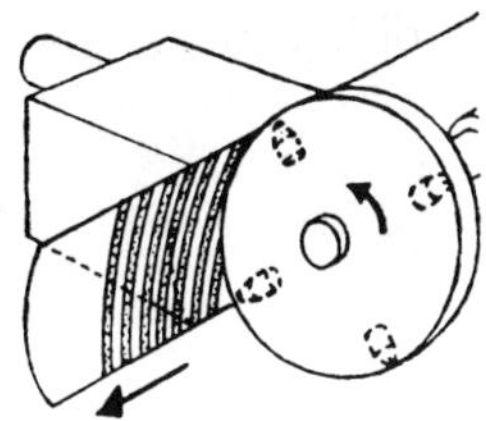

Figure Q.1 Quadruplex. The four heads produce transverse records on wide videotape

quantizing   Process in converting an analogue signal to digital form, in which it is sampled at regular intervals and the instantaneous amplitude expressed as one of a defined number of discrete levels, which can then be assigned a digital value; see *Figure Q.2*.

quantizing error   The error caused. by the limited number of discrete steps into which the input signal has been analysed.

quartercam   A *camcorder* system using ¼ inch tape videocassette. (Trade name).

quartz (crystal)   A slice of quartz cut to resonate at a stable defined frequency under the influence of an electric field using the ***piezoelectric*** effect.

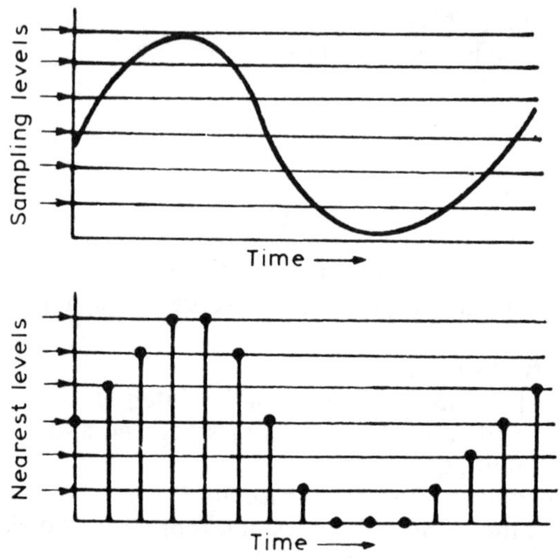

Figure Q.2 Quantizing

quartz iodine lamp A lamp in which the tungsten filament is enclosed in a quartz bulb containing an inert gas with a trace of a halogen, usually iodine. Also known as *quartz halogen lamp* and (less accurately) *quartz lamp*. See **tungsten halogen**.

qwerty The standard English typewriter keyboard layout, named from the first six characters of the top row of letters.

R

racking Another term for *framing*.

rack-over Movement to bring the viewfinder of a camera into line with its objective lens system.

radio frequency Frequencies generally above 20 kHz.

radio frequency interference Interference introduced into electronic circuits by radio transmitters or other sources of electromagnetic radiation.

radio microphone A microphone connected to a very small portable radio transmitter, the radiation from which is picked up by a local receiver, thus avoiding trailing wires.

RAI Radiotelevisione Italiana, the Italian state broadcasting network.

rails Portable light-weight tracks for smooth operation of a *dolly* on location.

RAM Random Access Memory. Memory that can have any *byte* of its data written in, or read out, in any order.

ramcorder Solid-state digital video recorder with Random Access Memory.

random access The ability to select at will any defined point from a magazine, programme or memory, as indicated by an address.

random noise Noise, or an electrical signal, that is random in its amplitude, frequency and phase characteristics.

random numbers Numbers not related to each other in any way.

raster (1) The pattern of horizontal lines forming the image scanning of a television system. (2) In computer graphics, a rectangular matrix of picture elements (*pixels*).

raster graphics A method of presenting graphical data as a display comprising groupings of *pixels* on the raster of a cathode ray tube. (cf. *vector graphics*)

raster unit The distance between the midpoints of two adjacent *pixels*.

raw stock Unexposed and undeveloped photographic film.

RC A Resistor/Capacitor combination. Often referred to as a time-constant expressed in seconds, the time-constant being the capacitance in farads multiplied by the resistance in ohms.

RDS Radio Data Service, additional coded data transmitted within the bandwidth of FM audio broadcast for automatic tuning, station identification, continuous time check, etc.

read To put information into a computer from a store, memory, disk or tape.

real In computer practice, a type of *variable* which is a precision number and may have a fractional or decimal component, e.g. TOTAL = 159.34 Also termed *floating point number*.

real time Keeping pace with events in the real world, as they are happening.

real-time programming Programming a *tape-slide* show while the tape is running at its normal speed.

rear projection Presentation of an image on a translucent screen by a projector placed on the far side from the viewer.

recce Reconnaissance or reconnoitre: a survey of a location.

receiver A device which receives data, particularly radio or television equipment which receives a modulated radio frequency and provides a demodulated output.

reciprocity failure Divergence from the photographic *reciprocity law* which may occur at extremely high or low values of light intensity or time.

reciprocity law In photography the relation which states that constant exposure will be obtained if greater intensity of light is compensated by proportionally shorter time and vice versa.

recovery time   Time taken for a device to return to normal after a signal above the processing threshold has been removed.

rectifier   A component in an electrical circuit which will pass electrical current in one direction but not in the other, hence a device which converts an alternating current into a direct current.

recursive   In computing, a process in which each step makes use of the results of earlier steps.

recursive filtering   Method of reducing video noise and random defects by the comparison of two or more adjacent frames, non-repeating items being eliminated.

redhead   Trade name for a portable mains-powered lighting unit, 800–1000 W, with adjustable beam for spot or flood.

reduction   Process of mixing signals from multitrack master tapes to produce a master tape for production; see also *mix down.*

reduction printing   Optically printing motion picture film to produce an image smaller than the original, usually on film of narrower gauge.

reel   (1) A flanged hub on which film or tape is wound. (UK film usage is *spool*). (2) A roll of film, the unit in which a programme or part of a programme is usually handled, either as the assembled negative or corresponding positive print.

reel band   Paper strip securing a tight-wound roll of motion picture print.

reel-to-reel   Film or tape path with separate feed and take-up reels in a camera, projector or tape recorder/reproducer, in contrast to the enclosed cassette or cartridge.

reflector   White or silver sheet used in location shooting to bounce natural light to illuminate shadow areas.

reflex camera   A camera in which the image in the viewfinder is obtained through the main objective lens, so that *parallax* error is eliminated.

reflex projection   A special effects process in which foreground action is photographed against a background image projected along the axis of the camera lens on to a highly reflective beaded surface. Also called *front axial projection.*

refraction   Deflection of a beam of radiation, such as light, on passing from one medium to another.

refractive index   Factor expressing the refraction caused by a medium; for light it is expressed as the ratio of the sine of the angle of incidence to the sine of the angle of refraction when the beam passes from vacuum into a medium; see *Figure R.1*.

refresh   (1) To recharge regularly the cells of a *volatile* digital memory so that their contents are not lost. (2) In computer graphics, the process in which the electron gun repeatedly draws a display on the surface of a *refresh tube.*

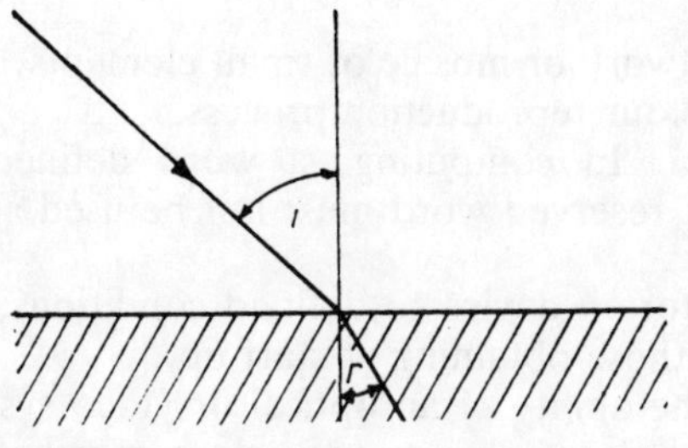

Figure R.1 Refractive index. The ratio
sin *i*/sin *r*

refresh cycle The period of one *refresh* of the display.

refresh tube A cathode ray tube in which the electron gun must be activated repeatedly to maintain the display.

register A short term store for digital information in a computer *central processing unit*.

register To place in exact alignment or position.

register bar A locating device consisting of a metal strip with two or more upstanding pegs, used to ensure accurate placing of material to be photographed or copied. The drawing or *cel* is edge perforated to match the pins.

register mount A slide mount containing locating pegs for precise location of the film by means of its perforations.

register pegs Pegs on a light box or rostrum, on which *cels* can be placed and kept in register.

register pins Locating pegs accurately shaped to fit into the edge perforations of film, paper or transparent sheet. In a camera, register pins precisely locate the film by means of its perforations.

registration (1) The precise location of successive images in a sequence so that unintentional movement is not introduced. (2) The correct positioning of three-colour component images to be co-incident without overlapping or fringing.

registration slides See *line-up slides*.

release print A motion picture positive print of picture and sound made for general distribution and exhibition.

relocatable In a computer, when *machine code* will execute correctly even if it is moved to a different address location.

remanence The residual magnetism of a ferromagnetic substance when, after being magnetised to saturation, the magnetising field is removed.

remote control Control of a device at a location remote from the device, by cable, radio, infra-red or other means.

reporter A commentator appearing in front of camera.

reseau A network or mosaic of small elements, as used in some forms of colour reproduction process.

reserved word In computing, a word defined for a special purpose. A reserved word must not be used as the name of a *variable*.

reset To restore a device to defined conditions, usually corresponding to those obtaining at start up.

resolution The ability of an optical or video system to produce separate images of objects very close together, and hence to reproduce fine detail.

resolving power The resolution of a system expressed in numerical terms.

resonance Sympathetic oscillation of a mechanical or electrical system at its own natural frequency.

retake The re-photography of a scene.

reticle Ground glass in a camera viewfinder engraved with exact size and position of camera gate, centre cross lines, etc. See *graticule*.

reticulation Distortion of a photographic emulsion surface into a pattern of small cracks and wrinkles, usually caused by severe temperature differences in the processing solutions.

retrieve To read or recover stored data.

retriggerable Describes a *monostable*, which can additionally be triggered to remain in its unstable (or abnormal) state for a further time period.

retrofocus lens An inverted telephoto lens design providing a long back-focus distance.

return See *carriage return*.

return to zero A method of digital magnetic recording in which the tape or disk is saturated in a positive direction for a 1 and in a negative direction for a 0, with a return to zero or no magnetising condition in between.

reveal See *conceal/reveal*.

reverberation The echoing of sound within a structure such as a room or building.

reverberation time The time taken for the echoes of a sound to drop 60 decibels in amplitude after the termination of the direct sound (60 dB is one millionth of its original value).

reversal film A film which after processing yields an image whose tonal distribution matches that to which it was exposed, e.g. a camera film giving a positive image, or a duplicating stock giving a negative image from a negative.

reverse To step back one slide with a slide projector or back one image with a dissolve pair.

revolve A turntable system.

rewind To transport a film or tape in the reverse direction, back onto its original hub or spool.

RF Radio Frequency.

RF cues In motion picture printing, cues applied to the edge of the negative in the form of small metal patches which are sensed by a radio-frequency detector.

RFI *radio frequency interference.*

RGB (1) Red Green Blue, the primary colours of an additive system in photographic or video colour reproduction. (2) Hence, a method of signal connecting giving direct access to the primary colours on a monitor or visual display unit.

RIAA curve Recording Industries Association of America recommended audio reproduce/equalisation characteristic, now universally adopted.

ribbon microphone A microphone operating by currents induced in a metal ribbon mounted in a magnetic field.

rifle microphone A highly directional microphone which can be aimed at its sound source.

rigging The setting up of lights, loudspeakers or projectors on a set, stage or venue.

ringing A damped oscillatory response at the beginning or end of a *pulse* signal. In a video display, it appears as closely spaced light and dark bands of decreasing intensity at the edge of an image where a marked change of brightness occurs; see *Figure R.2*.

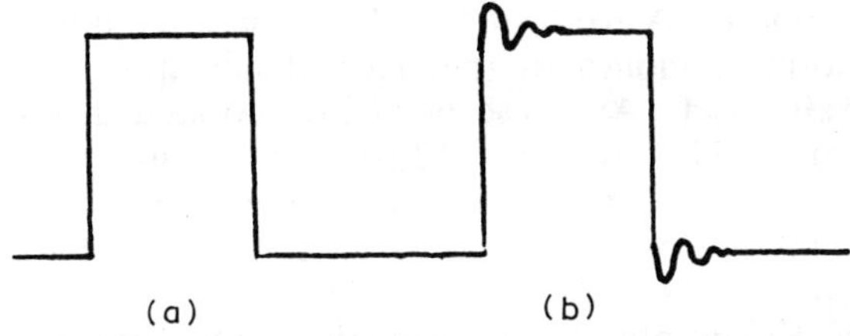

Figure R.2 Ringing. The input pulse (*a*) is distorted by oscillations at its leading and trailing edges (*b*)

ring modulator Device requiring two inputs to produce an output consisting of sum and difference frequencies between the signals. Used in electronic music synthesisers.

ripple (1) Unwanted AC signal superimposed on a DC supply. (2) A rapid succession of cuts or dissolves across a series of screen areas of a *multi-screen* presentation. (3) A *mix* effect in which the outgoing and in-coming scenes are distorted behind rippling lines.

rise time The time taken for the leading edge of a pulse to rise from 10% to 90% of its maximum amplitude.

risers The vertical steps of a *Fresnel lens*.

RMS *root mean square*.

RO Receive Only, a teleprinter term.

roam In a graphic display, to move the *window* through which a portion of the image is accessed and observed.

rock and roll The ability to move a video tape or synchronised picture and sound films backwards and forwards at varying speeds, to decide an edit point or for convenience in dubbing and mixing.

roll To start the camera(s), audio or video tape recorder(s).

roll feed Device to feed a roll of material into or across a machine. Specifically, the attachment to an overhead projector to feed the transparency material across the platen.

rolling title A series of titles or captions moving upwards across the picture area. (= *creeping title*).

ROM Read-Only Memory, used for storing non-varying data or instructions, e.g. *operating systems*.

root mean square Factor giving the effective value of an alternating current or voltage, taking account of its waveform.

roping Film damage in the form of continuous sprocket-tooth indentations, caused by *run-off*.

rostrum (1) Platform raising actors and scenery, often to camera height. (2) A camera stand designed to illuminate and hold in specified position the artwork to be filmed.

rostrum camera A fixed film or TV camera mounted vertically for shooting complicated graphics and animation.

rotary erase head An erase head incorporated in the rotating drum of a VTR; also termed *flying erase* head.

rotary printer A continuous *contact printing* machine in which both films are carried on a rotating sprocket at the time of exposure.

rotations Effects obtained by using the turntable on an animation stand.

rotoscope An instrument projecting a film frame by frame on to the table of an animation stand, for the preparation of animated drawings or as a background for animation *cels*. (Trade name).

rough animation The first tentative animation done as a test.

rough cut In editing, the first assembly of shots in their intended script order.

routine A name for a short stub program, which may be used several times.

RPS Royal Photographic Society (UK).

RS flip-flop A *bistable* which may be set or reset.

RS232   An *EIA* interchange specification containing a selection of *V24* interchange circuits. RS232 gives electrical characteristics and pin allocations in the 25-way connector.

RS422, RS423   Codings defining the control signals and types of connector required to provide a common interface between electronic units.   **RS422** is a differential 5 V system on two wires;   **RS423** allows for a high impedance state to allow more than one sending circuit to be on circuit.

RT60   The *reverberation time* for a 60 dB drop in amplitude.

RTBF   Radio-Télévision Belge de la communauté culturelle Française, Belgian state broadcast network in French language.

RTE   Radio Telefis Eireann, the Irish state broadcast network.

RTL   Register Transistor Logic.

RTP   Radio Televisão Portuguese, the Portuguese state broad cast network.

RTS   Royal Television Society (UK).

RTP   Radio Televisão Portuguese, the Portuguese state broad-cast network.

RTZ   *return to zero*.

rubber numbers   Numbers and other coding applied to processed rush prints and sound records for identification during film editing.

rumble   Very low frequency noise associated with audio disc turntables.

run   The instruction to commence execution of a programmed task, from the beginning unless otherwise specified.

run off   Misplacement of film so that it passes over the teeth of a sprocket, with resultant damage.

run out   (1) A length of blank film stock at the end of each reel to protect the film from damage. (2) The section of black film immediately following the last picture frame in the end section of an *Academy leader.*

run through   Rehearsal.

run time   Taking place when computer is run. (As opposed to when program is entered).

run up   The length of film or tape that has to run through a projector or audio or videotape recorder before it is operating smoothly at normal speed.

rushes   (1) In motion picture practice, the first prints, also termed dailies or rush prints, made from the newly processed picture or sound negative to check content and quality. (2) Unedited videotape from *ENG*.

RX   (1) Receiver. (2) Re-transfer, Re-transmit.

RZ   *return to zero.*

s second.

S *siemens*.

S100 A widely used 'standard' computer *bus* connection.

safe area The area within a frame that is certain to be in vision when transmitted on television or video.

safelight A source of visible light in a photographic darkroom whose colour and intensity allow unprocessed sensitive materials to be handled without danger of unwanted exposure.

safety base The almost non-flammable acetate or triacetate film base that has replaced the highly flammable nitrate base formerly used.

sample and hold A circuit element which samples the instantaneous amplitude of an electrical waveform and retains (holds) the value. Commonly used in analogue-to-digital converters.

sample print See *answer print*.

sandcastle pulse In electronics, a pulse able to convey different timing at different slicing levels. See *Figure S.1*.

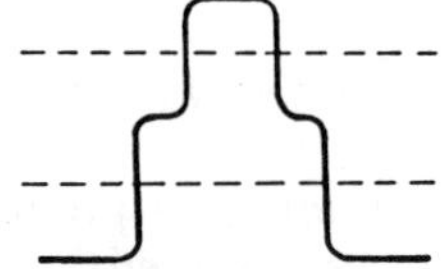

Figure S.1 Sandcastle pulse.

sandwiching Mounting two or more pieces of film in a single slide mount to create effects.

satellite (1) A sub-station of a main location. (2) A station in space, often geostationary, providing a radio-frequency link between ground-based transmitters and receivers.

Saticon TV camera tube of the Vidicon format with a photo-surface of doped selenium giving high resolution and reduced lag. (Trade name).

saturation (1) State when a magnetic material is fully magnetised. (2) In colour reproduction, the spectral purity of a colour.

save A computer instruction to store information on magnetic disk or tape. The term is also used for saving a program to bulk storage such as a core store.

sawtooth An electrical waveform which has a slow change in amplitude followed by a rapid return or vice versa; see *Figure S.2*.

SBCA Satellite Broadcasting and Communications Association (USA).

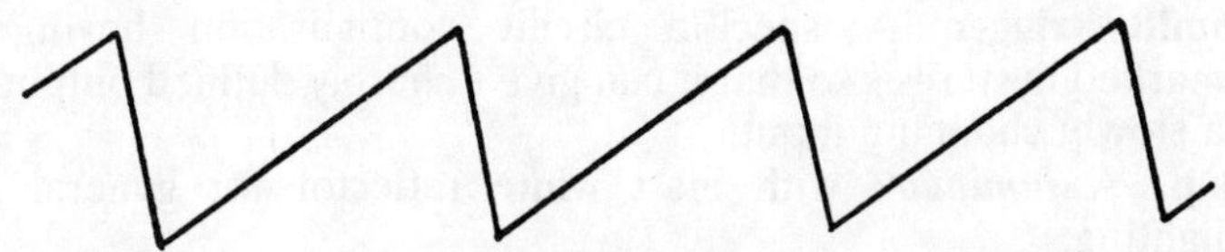

Figure S.2 Sawtooth.

SCA Subsidiary Communications Authorisation. Secondary channel used in stereo broadcasting in USA.

scalar In computing, a value or *variable* that is not part of an *array*.

scan line A horizontal line of *pixels* in a raster display.

scanner (1) Equipment used to carry out *scanning*. (2) Specifically, abbreviation for *caption scanner*. (3) Mobile video production vehicle.

scanning The process by which an area is systematically explored line-by-line, particularly in television, where an electron beam is used.

SCC Single Chip Controller.

scene The basic element of continuity sequence in a motion picture or television production, planned for recording as continuous uninterrupted action. Also, the action that takes place at one time and place, or its setting.

Scene Sync Trade name for a system which permits movement of the foreground camera synchronised with the mask camera in *chromakey*, and *colour separation overlay* special effects.

scientific notation Representing a large or small number as a power of ten, e.g. 1,252,000 as 1.252×10^6.

Schmidt optical system A large aperture optical projection system having a spherical concave mirror in combination with an aspheric correction lens; see *Figure S.3*.

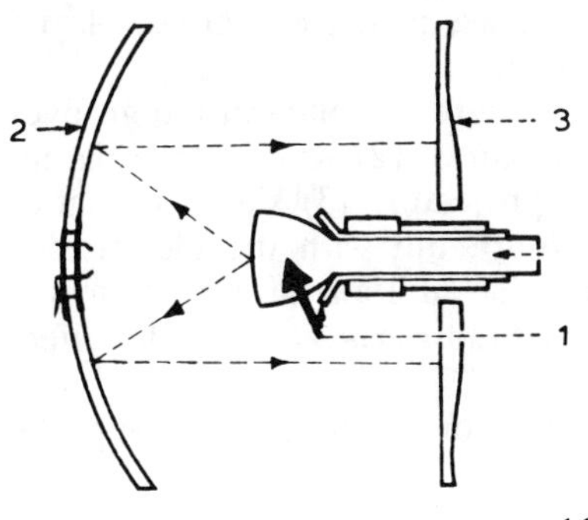

Figure S.3 Schmidt optical system. In projection television, the image on the CRT (1) is reflected by mirror (2) and corrected by the correction plate (3)

Schmitt trigger A specific circuit configuration having a marked hysteresis so that it can give a sharply defined output to a slowly changing input.

scoop A *luminaire* with matt white reflector for general *fill* lighting.

Scotchlite A highly reflective beaded screen material for *reflex projection*. (Trade name).

scrambling Continuous modification of a transmitted signal so that it can only be correctly received by use of a special device, the unscrambler.

scrape flutter A fault in magnetic recording resulting from stick/slip motion of the recording tape producing *flutter*.

scratchback Technique in which part of the artwork, painted on a *cel*, is removed frame by frame under the camera while the latter runs backwards.

scratch pad A temporary memory area used for holding partial results, etc.

screen (1) The surface on which a picture image is presented. (2) A conducting surface preventing external electrical or magnetic interference with a system.

screened cable A central conductor, or a number of conductors, insulated from an external screen formed by copper braiding or lapping or flexible metal.

screen dumping Copying the contents of a visual display unit screen to a printer or disk.

scrim Gauze used to diffuse light. Also metal gauze used to provide protection for bare bulbs.

script (1) The detailed scene-by-scene instructions for a film or television production, including description of the setting and action with dialogue and camera directions (= *scenario*). (2) A similar treatment for a tape-slide or multivision production. When the script also shows full details of the visuals, it is termed a *story board*.

scrix Video effect where the frame is divided into a number of rectangles which reduce and increase in size and are displayed in a different order on the screen.

scroll (1) Portion of disc surface where the pitch of the grooves has been increased to separate bands. (2) Roll of transparent film for use on an overhead projector. (3) Video transition effect similar to *push-on, push-off*, but with the picture displaced vertically. (4) In VDU the addition of a bottom line displacing all others upwards. With *spreadsheets*, a similar effect in any direction.

scrolling Adding a new line of information and moving the others vertically.

SCR Silicon Controlled Rectifier. A *thyristor*.

seamless masks See *soft-edged masks.*

search An ability to search rapidly backwards and forwards to other parts of the programme without loss of sound–picture synchronisation.

search time The time taken to complete a search and find the required item or part of the programme.

SECAM Sequential Couleur à Mémoire. Sequential colour with memory. French television system.

secondary battery A battery composed of *secondary cells.*

secondary cell An electro-chemical cell which has to be recharged from a prime source before further electrical energy can be drawn from it.

secondary (transformer) A transformer winding other than the *primary*.

segment In computer graphics, a number of display items which can be manipulated as a single unit.

segue An indication that one section of music is to be played immediately after another one; this may be straight cut or cross-fade.

Selectavision Trade name for a grooved capacitance videodisc system.

self-blimped Describing a motion picture camera whose operating noise level is so low that no additional sound-proof enclosure is necessary.

self clock A data system requiring only one channel for data and the clock timing needed to receive that data.

sell-through Distribution of *videogram* by direct retail sale to the public rather than by rental.

Selsyn Trade name for a servo system.

sensitometry The scientific study and measurement of the effect of light on photographic materials, especially the relation between exposure and the resultant density after processing.

separation Acoustic isolation between instruments essential to a certain degree in multitrack recording to give control over relative balance between instruments.

separations See *colour separations.*

sepmag A magnetic sound record, on tape or perforated film, separate from the picture film with which it is associated.

sequential access A mode of data retrieval where each byte of data is recovered in the order in which it was written to the disk.

sequential scan TV *scanning* system in which all lines comprising a frame are transmitted in sequence rather than *interlaced*.

serial communication A data communication scheme needing only one wire (and return) where data is sent one bit at a time.

servo A control system involving feedback, which operates to correct any error between the input and an output sensor.

set (1) The studio floor area where the action is to be played, including the scenery. (2) To place a *bistable* in the '1' state.

set level A continuous tone at exactly signalling level, used to adjust recording level. It is used in multivision control systems.

set up (1) The complete system. (2) In motion picture production, the arrangement of setting, actors, lights, microphones and cameras ready to record a scene; also, a specific camera position. (3) To get into a state of readiness to operate. (4) In a TV system, to differentiate between zero level or true black, and the actual black level of the reproduced picture.

SFD Society of Film Distributors (UK).

SFX Sound effects.

shadowboard A black board suspended below the camera on an animation stand, containing a hole through which the lens protrudes to prevent reflection of the camera in the platen glass.

shadowmask A display tube for colour TV having a mosaic of *RGB* phosphors to which the electron beam is directed through a perforated metal screen; see *Figure S.4*.

shedding Magnetic recording material losing particles from its magnetic coating.

SHF Super High Frequency, 3 GHz–30 GHz.

shielding In computer graphics, defining an opaque *window* in which to display a message.

shift register A series of *bistables* capable of storing and transmitting a pattern of ones and zeros, one bit at a time at each *clock pulse*.

shipping reel A heavy-duty 2000 ft spool used for film release print distribution in the United States.

shoot To operate a camera.

short end The portion of a roll of film or magnetic tape remaining after the main part has been used.

shot A scene photographed or recorded as one continuous action.

shotgun microphone A highly directional microphone. See *rifle microphone*.

show copy A selected copy of a completed programme, video, tape-slide, multivision or film, intended for presentation to an audience. For a motion picture production it is often termed a *show print*.

Showscan Trade name for a high-quality cinema presentation system using 35 mm film projected at 60 frames per second.

shunt (1) To connect two components in parallel, e.g. a resistor placed across an ammeter to increase its range. (2) To place a load across terminals.

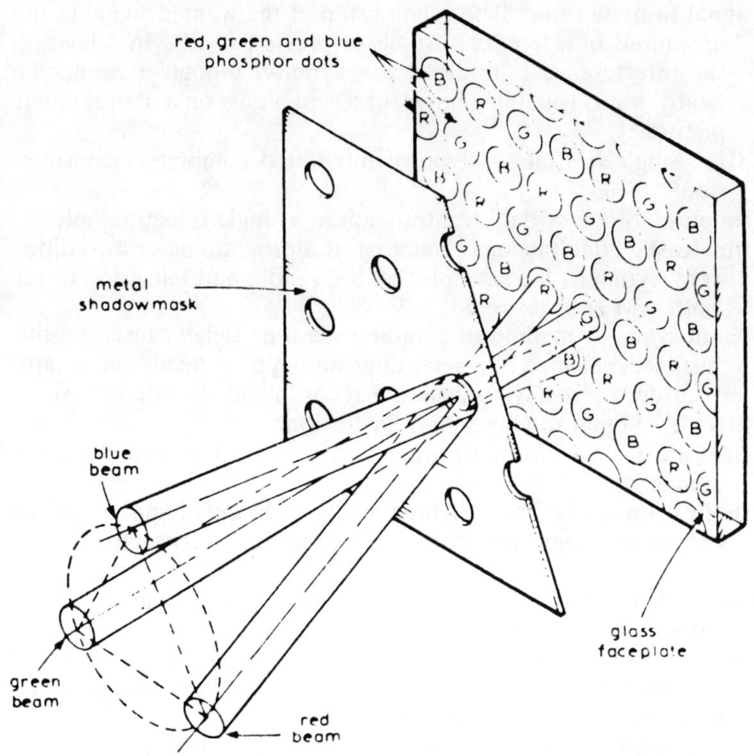

Figure S.4 Shadowmask. The screen is covered with primary-colour phosphor dots in the form of triads for each colour cell, and a metal mask ensures that only the beam with the correct colour signal modulation strikes each dot.

shutter A device to cut off light in an optical instrument, particularly (1) In a camera where it is opened and closed to effect the exposure. (2) In motion picture equipment, a rotating blade which interrupts the light in a camera or projector while the film pull-down occurs. (3) In slide projection, a device to cut off the light while the slide is being changed or if there is no slide in position. See also *snap*.

shuttle search The ability to play back a film, video- or audio-tape in either direction and reproduce the picture over a wide range of speeds.

SIAD Society of Industrial Artists and Designers.

sideband A range of frequencies generated by modulating a carrier wave.

siemens Unit of electrical conductivity, the reciprocal of an ohm, equivalent to *mho*.

signal-to-noise ratio   The relationship of the wanted signal to the unwanted interference, usually expressed in dB. In television the interference is referred to as 'noise' though it cannot be heard, but is roughly equivalent to graininess on a photographic picture.

SIL   Single-In-Line. A type of integrated component construction.

simplex   Mode of data transmission in a single direction only.

simulcast   Simultaneous broadcast of a programme on two different channels, for example, on both radio and television or on both AM and FM radio.

Simul-sync   A method of avoiding the time delay caused by the distance between the recording and replay heads on a tape recorder by replaying from the record head. (Trade name).

SINAD   SIgnal to Noise And Distortion.

sine wave   Waveform showing variation in a simple harmonic manner.

single frame   (1) One individual image in a strip of images. (2) To expose or project one picture at a time at a comparatively slow rate.

single shot   Exposing film, one frame at a time, as in animation work.

single standard   A TV monitor, video tape recorder or other equipment, capable of accepting video signals of only one standard, *NTSC*, *PAL* or *SECAM*.

single system   Method of motion picture production in which both picture and sound are simultaneously recorded on the same strip of film.

sit   Depression of a video signal towards black, compressing darker tones.

skew   Picture distortion in which the verticals are not at right angles to the horizontals. In video recording it can result from mechanical error in the tape motion or tension. The term also refers to the non-simultaneous arrival of data on several data lines.

skip frame   In cinematography, printing only frames selected at regular intervals from the original record to produce the effect of speeding up action.

skivings   Fine hair-like slivers of film removed from its edge by a mechanical obstacle during running.

skypan   A large *luminaire* with matt white reflector for lighting back drops.

slant azimuth   In magnetic recording, setting the head gap at an angle other than 90° to the direction of tape movement; different slant azimuths on adjacent tracks avoid *crosstalk* without *guard bands*. See *Figure S.5*.

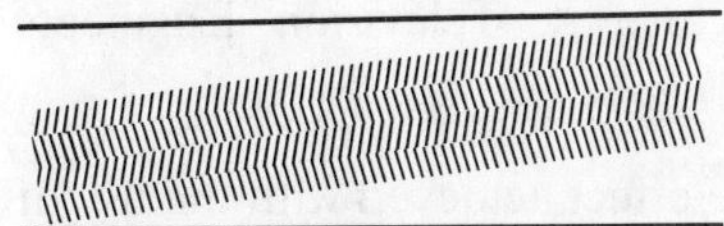

Figure S.5 Slant azimuth. The azimuth angle differs on adjacent tracks to avoid cross-talk.

slave A unit designed to function only as ordered by a 'master' unit, e.g. a videotape recorder that is controlled by another VTR.

slate Chart or board with information regarding the production, its title, scene and take numbers, etc., reproduced at the beginning or end of each *take* photographed on film or recorded on tape.

slew rate Rate of change of voltage.

slide The complete assembly of a still picture transparency in a mount, which may also contain a mask.

slide mount A metal or plastic receptacle, often glazed, for holding a picture image on film.

slide-tape See *tape-slide*.

slider A form of *potentiometer*.

sliding frequency Also known as *gliding frequency*.

slit In photographic sound, a narrow aperture through which the film is exposed in recording and scanned in reproduction.

slit loss In photographic sound, the reduction in amplitude at higher frequencies which results from the finite height of the slit in recorder or reproducer. Note that slit height is measured in the direction of film travel.

slope Steepness of filter curves normally expressed in dB/octave.

slot loading Mechanism where tape or film can be loaded laterally instead of lacing or threading.

slo-mo Abbreviation for *slow motion*.

slot mask tube *Shadowmask CRT* in which the mask and phosphors are in vertical lines rather than the usual triads of dots.

slow motion Photography by a motion picture camera with the film running faster than normal, so that when the result is projected at normal speed, the action appears to be slowed down.

slow scan TV scanning and transmission at a slower frame rate than normal, used in *video conferencing* to reduce the required bandwidth of the system.

smart receiver Jargon for a receiver presenting *HDTV* in a wide-screen format.

SMATV Satellite Master Antenna Television, cable or microwave distribution of satellite broadcasts to individual viewers from a central antenna.

SMPTE Society of Motion Picture & Television Engineers (USA).

S/N *Signal-to-Noise (ratio).*

snap A very rapid picture change effect achieved with the use of shutters in slide projectors.

SNG Satellite News Gathering, usually linked to its base by *Ku-band* transmission via a satellite.

snoot Conical hood to reduce the width of the beam from a light source.

snow Random *noise* or interference appearing in a video picture as white specks.

SNV Satellite News Vehicle, self-contained unit for *SNG*.

SOF Sound-On-Film. Term used to indicate synchronised picture and sound on film.

soft edge A diffuse or graduated boundary to a picture image area, as in *matte* or *wipe* effects.

soft-edged masks Graduated neutral density masks used in slide mounts to blend adjacent projected images imperceptibly so that very large composite images can be made without visible joins.

soft focus Producing an image with less than the maximum sharpness of which the system is capable.

softlight *Luminaire* with an open bulb in a matt white reflector giving substantially shadowless illumination.

software (1) The program of instructions which the *hardware* of a computer obeys. (2) The audio and visual programme material used in A-V productions.

solarisation Originally, a photographic effect in which the picture image is partially reversed in tone with light or dark edges at highlight and shadow boundaries; a similar effect produced by video methods.

solid state Not using liquids or gases, hence semiconductor technology generally.

sort To place in a specified order, usually alpha-numeric.

sound drum Roller used to control uniformity of movement of film where a scanning-beam slit scans, or a magnetic head reads, the sound track.

sound gate The gate used instead of a sound drum to keep the film sound track aligned with the scanning beam.

sound head (1) In motion picture equipment, a mechanism for transporting the film and reproducing the photographic or magnetic sound record. (2) In audio or video equipment, the mechanism for recording/reproducing the audio information.

sound track The defined area along the length of a recording medium such as film or magnetic tape which carries the audio information. Also, the sound record itself.

source code A program written in a high-level language; needs *compiling* or *assembling* before use.

spacing loss In magnetic recording, the loss of high frequencies due to imperfect contact between the reproducing head and the magnetic medium.

sparks An electrician.

speaker support module Slide sequence for use with a live presenter.

special effects General term for scenes in which an illusion of the action required is created by the use of special equipment and processes rather than in reality. Also, video mixer enabling sections from two or more pictures to be montaged and displayed on the same screen.

spectrum (1) A range of wavelengths or frequencies of acoustic, visible or electromagnetic radiation. (2) Specifically a display of visible light radiations arranged in order of wavelength.

spectrum analyser Device to display the frequency distribution and levels of a signal.

speech track A voice track as opposed to music or effects.

SPG Sync-Pulse Generator.

spherical aberration Lens aberration in which rays through the peripheral portion of a lens come to focus at a point different from the axial rays; see *Figure S.6*.

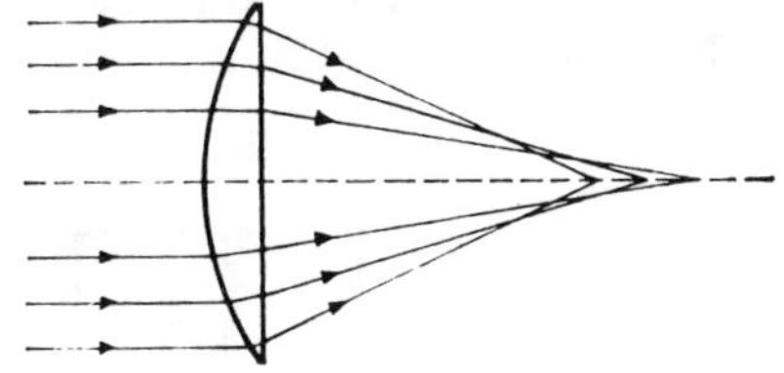

Figure S.6 Spherical aberration. The point of focus differs for the outer and inner zones of the lens

spider Three-armed *spreader* for *tripod* legs.

spigot An adaptor screwed to the hanging bolt of a *luminaire* case to enable it to be mounted on a floor-stand.

spike A short pulse superimposed on another electrical signal, e.g. a switching *transient* carried through the mains AC supply, as opposed to a *glitch*, which is a short unwanted signal not carried on another.

SPL Sound Pressure Level.

splice A physical join in the medium, film or tape.

split edit A video editing change where sound and picture cuts occur at different times.

split screen A shot in which two or more separate images appear in different areas of the same picture.

spoilers   Pins used to prevent head-to-tape contact during rewind or fast-forward.

spoking   Distortion in a reel of film caused by loosely winding badly curled material.

spool   A flanged hub on which film, magnetic tape or ribbon is wound.

spool, to   To wind a film or tape at a speed higher than the normal speed of reproduction.

spotting   (1) Locating individual sounds or words in a sound recording. (2) Retouching with an opaque medium to eliminate small transparent defects, such as pinholes, in the heavy density areas of a photographic image. (3) Marking slides to indicate their correct orientation for projection.

spot wobble   Vertical oscillation of the scanning beam in a television display to render the raster spacing less obvious.

spreader   A triangular or Y-shaped device used on the floor to fix the legs of a *tripod*.

spreadsheet   The computer equivalent of accounting paper, able to calculate in columns and rows, used mainly for financial statements.

sprocket   A toothed drum engaging with the perforation holes of film or paper tape for transport.

sprocket holes   Deprecated term for *perforations* in motion picture film.

square wave   Electrical waveform switching virtually instanteously between two voltage levels, with the 'on-off ratio' not necessarily of equal duration; see *Figure S.7*.

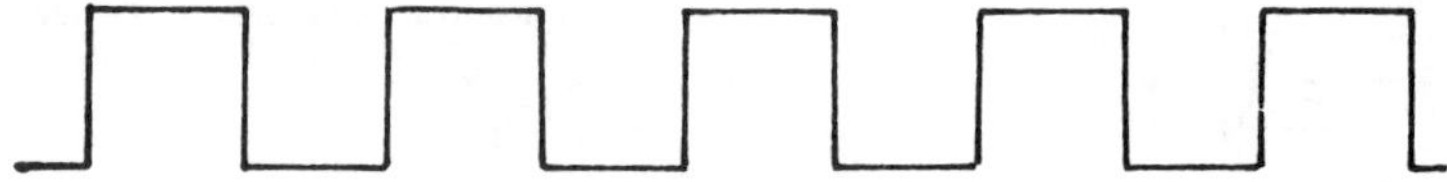

Figure S.7 Square wave.

squeezezoom   Device to manipulate the geometry of a television image for artistic effect; specifically a trade name, but loosely used for other equipment to perform similar functions.

squegg   An unwanted oscillation which leads to instability.

squelch   A circuit in a FM radio receiver which *mutes* the output when the incoming radio signal falls below a preset amplitude.

SSVR   Solid State Video Recorder, using semi-conductor storage in place of magnetic tape or disc.

stack   A store in a microcomputer for holding program count or data in a first in, last out manner, used to ensure orderly return from *nested subroutines*.

stacker   See *projector stack*.

stage left   On the actor's left looking at the audience.

stage right   On the actor's right looking at the audience.

stage weight   Weight used to keep french scenery braces steady.

staircase   A waveform which ascends or descends in steps between set limits.

standard cell   A primary cell characterised by its constancy of *EMF* over a long period of time, used as a voltage reference.

standards converter   Equipment to convert television pictures from one scanning standard to another, typically US *NTSC EIA* to European *PAL CCIR.*

standby   A waiting state, usually with reduced power or lamps off.

star filter   A filter with an engraved or etched line pattern to produce star effects on bright light sources.

star wheel   The slotted '*maltese cross*' wheel in a *geneva* movement.

start bit   In serial data transmission, the first element in serial communication, preceding the first data bit.

static   (1) Interference (noise) in sound or television. (2) Strictly Electrostatic, high voltages which can be generated by friction and may damage electronic components, particularly semiconductors.

static marks   Marks caused by discharge of static electricity on or near the surface of undeveloped photographic material, visible after processing.

static memory   A semiconductor memory where the information is retained in the absence of refreshing or recirculating.

statement   A computer term for a single program instruction such as 'LET $X = Y + 10$'.

STC   Single Time Code, in videotape recording.

Steadicam   Trade name for an individual harness with gyroscopic camera mounting, providing smooth action with a hand-held camera.

step   To proceed through a computer program manually one instruction at a time.

step printing   Motion picture printing in which the film is exposed intermittently, frame by frame.

stepper motor   A type of motor which rotates in small but fixed steps, at any of which its position can be held.

stereo   Abbreviation for stereophonic or stereoscopic.

stereophony   Sound reproduction using two or more channels to give the impression of spatial distribution.

stereoscopy   A system of photography giving the viewer a three-dimensional effect; it may use two separate images filmed from separate positions, one to be viewed with each eye.

stethoset   Type of headphone like a doctor's stethoscope.

still frame   Continuous reproduction of one stationary image from a train of recorded images, in either motion picture or video.

stock   General term for unexposed cinematograph film. (*Raw stock*).

stock numbers   Sequential identification numbers applied at fixed intervals on the edge of cinematograph film during original manufacture.

stock shot   = *library shot*; see **library**.

stop bit(s)   In serial data transmission, the period or periods of equal duration to the data element which terminate a character or block. They ensure enough time to be ready for the next *start bit*.

stop frame   In cinematography, the repeated printing of a single frame to appear as a still picture when projected.

stop   The opening of a lens controlling the amount of light transmitted (= *diaphragm*).

stop motion   The operation of a motion picture camera, printer or projector one frame at a time.

store   Device or system into which data can be entered for retrieval later.

storage tube   A cathode ray tube which does not require *refreshing* to maintain its display.

story board   Chart consisting of a series of still pictures, usually drawings, summarising the contents of a proposed film, video or slide production.

streaming tape drive   A type of digital data recorder for bulk storage, used for *dumping*, usually recording successive tracks continuously in alternate directions.

strike   (1) To dismantle and remove *sets* or *props*. (2) To light an arc lamp.

string   In computer practice, a type of *variable* consisting of a sequence of characters, correctly referred to as a literal, enclosed in quotation marks, e.g. DAY = "TUESDAY".

stringy floppy   In computer data storage, a bulk storage system based on magnetic tape, usually endless.

stripe   A narrow band of magnetic material applied to photographic motion picture film for sound recording and reproduction.

stripe mask   The mask of a *slot mask tube*.

striping   (1) Applying a magnetic stripe to motion picture film. (2) Adding longitudinal *time-code* to video rushes after shooting.

strobe lighting   Electronic flash lighting synchronised with a motion picture camera frame rate, giving sharp images of fast moving objects.

strobing   A disturbing effect in film or television arising from the field or frame sampling rate distorting motion, particularly of rotating objects.

stroboscope   (1) Pulsed light source used to measure speed of rotation or frequency of repetition or to freeze the motion of objects. (2) A disc with equally spaced marks around its circumference used to check the rotational speed of objects, such as checking tape or turntable speed in the presence of AC lighting.

STV   (1) Subscription Television. (2) Sverige Television, subsidiary of the Swedish national broadcasting corporation, Sveriges Radio.

stylus   (1) Needle for disc reproduction, usually replaceable, may be made from sapphire or diamond. (2) In computer graphics, a pencil-like device for the input of information from a *data tablet* or for selection from a *menu*.

subcarrier   A band of frequencies superimposed on to a main carrier frequency. In colour TV waveform, *chroma* is carried on a modulated subcarrier, which is superimposed on to the upper frequencies of the *luminance* signal.

subroutine   A section of program to do a specified task for a main program or for another subroutine. Subroutines may be called several times for different parts of the program.

subtitle   A line or lines of words superimposed at the bottom of a film or video picture, translating foreign dialogue or as an aid to the hard-of-hearing.

super   *Superimpose.*

superimpose   To add one picture on top of another, usually so that both continue to be visible; to add a caption or graphic over a picture.

superslide   A slide with a large image area, 40 × 40 mm, in a standard 50 mm square mount.

supertrouper   A very large follow spotlight.

suppression   The reduction of generation of RF interference, e.g. by fitting a resistor-capacitor circuit to contacts.

surface wave device   A device (such as a filter) where the operation depends on an acoustic wave in the surface of the device.

surround-sound   Sound reproduction system with a quadraphonic arrangement of loudspeakers providing sound from 360°.

S-VHS   Super-VHS, a higher quality modification of the *VHS* system.

SVR   Super Video Recorder. (Trade name).

sweetening   (1) The compilation of complex sound effects to match the visual images. (2) In video, enhancement of a recorded image by electronic modification of tonal and colour rendering, edge sharpening, *gamma* correction, visual *noise* reduction, etc.

SWG Standard Wire Gauge.

switched star A form of cable TV distribution.

switcher (1) Switching Mode Power Supply (slang). (2) American term for vision mixer.

sync Synchronism or Synchronisation.

synchronisation The fitting together in accurate time-relationship.

synchroniser Group of two or more sprockets on a common shaft to allow lengths of perforated film to be wound through in a fixed relationship.

synchroniser, frame Device to store a full frame of video information which may be read out at a different rate to the input to provide a synchronous output.

sync pulse (1) In motion picture practice, a signal directly related to the speed of the camera, recorded on the magnetic audio tape for subsequent synchronisation. (2) The part of a composite video signal that controls the repetition rate of the scanning system; see *Figure B.1*.

synthesiser (1) An electronic musical instrument used for producing sound effects or electronic music. (2) An electronic circuit which can accurately produce several different frequencies from a single reference frequency.

T

T Tesla. Unit of magnetic induction.

T Tera-, a prefix denoting a factor of 10^{12}.

T-number A system of calibration of the light transmission of a lens based on its actual transmission at various diaphragm settings. (cf. *f-number*).

tablet See *data tablet*.

tabs (1) In computing, tabulation, setting printout or text on a printer or visual display unit in vertical columns. (2) Markers, often metal clips, attached to the edge of motion picture film to identify or cue a position. (3) Openable curtains on a stage. (US term is *drapes*).

TAC Television Advisory Committee (UK).

tachometer A transducer which converts speed of rotation into a scale reading or an electrical signal related to speed.

tail The final section of a roll of film or tape.

take A scene or part of a scene photographed without a break in the action.

take-up That part of a machine handling film or tape at which material is wound up after passing through.

talkback Communication between director and technical and production staff in studios.

tallylight The cue light on a TV camera.

tape Storage or recording medium in the form of a long narrow strip, either a magnetic-coated plastic base or a paper strip with punched holes.

tape-slide A sequence of slides synchronised with, or controlled by, an accompanying audio tape.

tape splice A join in film or tape in which the ends are joined by adhesive tape.

tape transport Mechanism for driving and guiding audio or video magnetic tape in a recorder/reproducer.

target The surface of a video camera tube on which the optical image is formed and scanned.

tarif Technical Apparatus for the Rectification of Indifferent Film: equipment for adjusting colour reproduction when film is reproduced on a *telecine* system. See also *masking (video)*.

TASC Television Advisory Technical Sub-Committee (UK).

TBC *time-base corrector*.

TCIP Time Code In Picture, visible *time-code* numerals in videotape editing.

TDM Time Division Multiplex.

TDS Time Delay Spectrometry.

TEA Theatre Equipment Association (USA).

teaching wall An integration of facilities such as flipcharts and white-boards with a front, rear or OHP screen.

telecine Equipment for replaying motion picture film into a television system. See also *flying spot scanner*.

Telecom The British national telecommunications service.

teleconferencing Form of *video-conferencing* in which the picture is only updated at intervals, to reduce the transmission band-width required.

telephoto lens Camera objective lens of long focal length and comparatively short back focus.

teleprinter Keyboard and hard-copy print-out for transmitting and receiving alphanumerics over a communications system.

teleprompter Television prompter providing the script for an artist to read while looking directly at the camera. (Trade name).

telerecording Transferring a television or video presentation to motion picture film.

Teletel French viewdata service.

Teletex An intercommunicating system between text-only terminals, such as telex, not involving video display. (Trade name).

teletext A *videotex* data service, transmitting alphanumerics and graphics in the blanking interval of broadcast TV signals.

teletype A teletypewriter, a term for *teleprinter.*

television System of transmitting electrical signals for the reproduction of a visible image at a distance, often with associated sound.

telex System of international teleprinter communication over public telegraphy links.

Telidon Canadian viewdata service.

Telset Finnish viewdata service.

tentelometer Trade name for a tape tension measuring instrument.

tera- A prefix denoting a factor of 10^{12}.

terminal A device remote from the computer, at which data can enter or leave a network.

terminator A load inserted at the end of a transmission line to prevent the signal from bouncing back.

tesla Unit of magnetic induction.

test tape A pre-recorded magnetic tape for the alignment and testing of tape recorders/reproducers.

TF 1 Télévision Française 1, the first French state broadcast TV network, now privatised.

thaw The return to action after a *freeze frame* effect, especially in videowall presentation.

THD (1) Total Harmonic Distortion. (2) Third Harmonic Distortion.

theatre/theater In American usage, 'theatre' tends to be used for a cinema, 'theater' for the live stage.

thermal printing (1) A non-striking system using a print head with a number of quick-heating elements, operating with heat-sensitive paper or, using a special ribbon, with ordinary paper. (2) Also, an imaging system using encapsulated bubbles in an emulsion.

thermal trip A protective device which mechanically interrupts the circuit if it detects over-heating.

thread See *lace*.

3-D Three-dimensional. (1) A stereoscopic system which provides separate left-eye and right-eye images to the viewer, giving the appearance of solid objects in space. (2) But in video and computer graphics, the term merely indicates the use of perspective and gradation to represent solid objects on a normal screen.

three-perf System of 35 mm motion picture photography using a frame 3 perforations high instead of the standard 4 perforations, thus reducing the length of film used for a given time by 25%.

three phase A three (or four) wire means of distributing AC electricity power supplies with the waveform in each of the three wires 120° out of phase. The fourth wire is the neutral in a star system.

three-plane registration In slide projectors, the exact positioning of each slide mount in three dimensions, vertical, horizontal and along the optical axis.

threshold howl Acoustic feedback produced by sound from loudspeakers re-entering microphone.

throw The distance from the projector aperture to the centre of the screen.

thumbwheel A rim-operated rotary control; in computer graphics it controls the movement of a line, either horizontally or vertically, across the display surface.

thyristor A controlled semiconductor switch often used as a power control device, e.g. in lighting dimmers.

Tic-tac Viewdata type system used in France.

tie line An interlinking sound or vision connection routed between technical areas.

tile Video effect in which the image is divided into rectangles of adjustable size, within which colour and brightness are integrated and reproduced as an average value.

tilt (1) Rotation of a camera in a vertical plane. (In contrast to *pan*) (2) Changing the slope of the frequency response in a sound reproducer.

TIM Transient Intermodulation Distortion.

timebase Electronic circuit which causes the spot to scan a cathode ray tube to form the raster. Each line is scanned at constant speed.

timebase corrector Device to correct timing errors in video signals, usually originated by deviations in speed in video tape recorders.

time code Coding system, usually binary, recorded on audio and video-tape, and sometimes on film, for subsequent synchronisation and editing. It denotes hours, minutes, seconds and allows unique identification of frames.

time lapse Recording a sequence of images with a regulated (and sometimes lengthy) time delay between each picture, using either film or video.

timer A mechanical, electromechanical or electronic device for controlling the time or sequence of events.

time shift   The practice of recording a television broadcast programme for replay at a later time.

tip projection   Penetration depth of video head into the video
tape. Measured as the protrusion of the video head outside the
drum.

titles   Words appearing in a motion picture or television production which do not form part of a scene.

TK   *telecine.*

TMD   Thermal Magnetic Duplication, a system of high-speed
duplication od videotapes from a mirror-image master.

toggle   A *bistable*. Also, to clock from one stable state to the
other.

tone   (1) The variation in a colour, or in the range of greys
between black and white. (2) A constant audio frequency.

tone arm   See *pick-up arm.*

tone control   Electronic circuit for manually modifying the frequency response of amplifiers, commonly effecting boost or cut
in treble and/or bass.

top hat   A small camera mount used when a very low position is
required. (= *hi-hat*).

TOPIC   Teletext Output of Price Information by Computer.

toroidal transformer   High efficiency transformer characterised
by circular shape and immunity to, and low production of,
external magnetic fields.

touch button   A button which operates by capacitance or other
effect, without requiring contact closure.

touch screen   Visual display from which an item may be selected
by physically touching that part of the screen.

track   A defined part of the width of recording medium, photographic or magnetic, carrying discrete information.

tracker ball   A two-dimensional control in the form of a rotating
ball in a socket, used in computer graphics to control the
position of a cursor.

tracking   (1) Physical movement of a camera and its mount
towards or away from the subject, or to follow a moving action.
(2) In electronics, a narrow band electronic filter which automatically alters its centre frequency according to the frequency
of the incoming tone. May be used to track (follow) the
fundamental tone or its harmonics. (3) In disc replay, the path
of the pick-up head following the grooves or other method of
location. (4) In video tape recording, causing the video heads to
follow exactly the recorded video tracks.

trailer   (1) A short film advertising a forthcoming cinema presentation or a similar technique in television. (2) The identification and protective leader at the end of a reel of motion picture
film.

tranny   Abbreviation for transparency or a transformer.

transcoder   System to convert one TV colour standard to another, e.g. *PAL* to *SECAM*.

transcription   (1) A written text from a sound track. (2) A term for a high-quality audio disc turntable.

transducer   Device for converting electrical, acoustic or mechanical energy into another of these forms.

transient   Momentary disturbance to an electronic signal.

transmitter   A device which sends information. It may be a transducer, radio or television system generating and distributing signals.

transmission   That information sent by a *transmitter.*

transparency   An image to be viewed by transmitted light, either directly or by projection.

transparency viewer   Device containing diffuser and lens (sometimes with a self-contained light source) for viewing a transparency.

transparent   (1) Allowing the passage of light or other energy without distortion and diminution. (2) Hence in video, describing a condition where the output signal is identical to the input, without distortion being introduced. (3) In computers, a condition where the action of part of a system is effectively undetectable to the person or system using it.

transponder   A transmitting/receiving system which will respond with its own signal when interrogated.

transposer   A television repeater that does not fully demodulate the received signal before transmitting it.

transverse scan   Method of video tape scanning in which the tape is curved across its width and scanned by one or more heads rotating at 90° to the plane of tape motion; see *Figure T.1.*

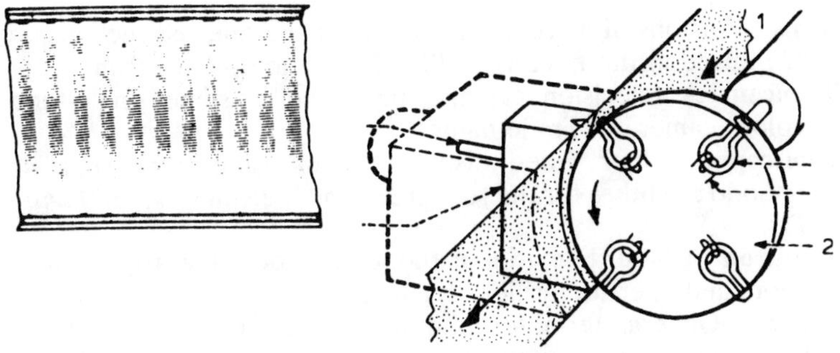

Figure T.1 Transverse scan. The wide videotape (1) is scanned by four heads on the rotating drum (2)

trapezium distortion Image distortion in an optical or video system causing a rectangle to appear with one or more sides converging instead of parallel. cf. *keystone distortion*. See *Figure T.2*.

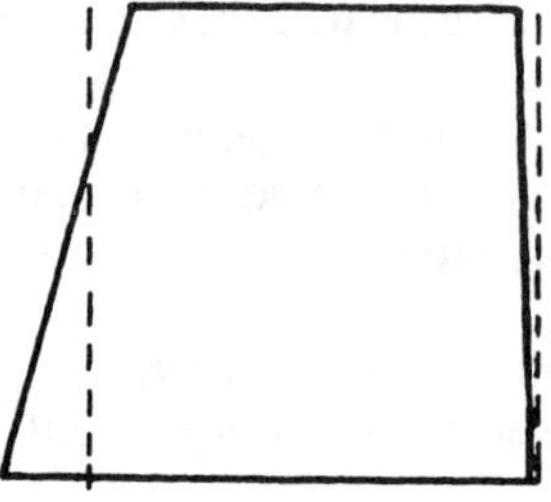

Figure T.2 Trapezium distortion

travelling matte In cinematography, a trick shot in which the foreground action, usually photographed against a blue backing, is superimposed on a separately recorded background by laboratory printing processes using *mattes*.

travelling peg bars Peg bars recessed into an animation table which can be moved mechanically in an east/west direction.

tray American term for a slide magazine, typically a horizontal rotary type.

tree and branch A form of cable distribution system.

triac A bi-directional version of the *thyristor*.

triangle Three-sided *spreader* for *tripod* legs.

triax A coaxial cable with an extra screen capable of transmitting power and coded commands to a television camera as well as receiving the video signal from the camera. Used when the camera is a long distance from the control unit.

trigger (1) To initiate an action. (2) Mechanical lever to start or stop an action. (3) Electronic pulse of short duration to operate a circuit.

trims Portions of a scene left over after the selected section has been used in the final assembly of a motion picture film.

Trinicon A television camera tube used in some single-tube colour cameras. (Trade name).

triniscope Colour video display system using three separate cathode ray tubes for the red, green and blue images; see *Figure T.3*.

Trinitron Colour television cathode ray tube using striped phosphors and aperture grille. (Trade name).

tripack General term for photographic material having three layers of sensitive emulsions for recording and reproducing colour images.

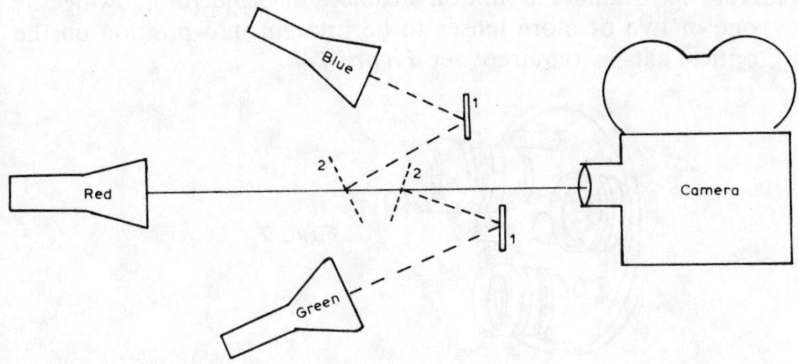

Figure T.3 Triniscope. Separate red, green and blue video images on three tubes are combined by the reflectors 1, 1 and the dichroic mirrors 2,2 for photography on film in the camera.

triple standard A TV monitor, video tape recorder or other equipment, capable of accepting video signals of *NTSC*, *PAL* or *SECAM* standard.

tripod An equipment support for a camera, screen, etc., having three legs adjustable in height and spread.

tristate Electronic circuit having three states, a low impedance 1 and a low impedance 0 state, and also a high impedance condition to permit other devices to signal on the same wire.

truck See *tracking*.

TTL (1) Through The Lens, a camera viewfinder system. (2) Transistor-Transistor Logic. A common low cost integrated circuit logic family operating from 5 V.

TTY Teleprinter (Teletype, Telewriter).

tumbling In computer graphics, moving a three-dimensional object on a display by continually changing its axis of rotation.

tungsten-halogen lamp An efficient lamp consisting of a tungsten filament in a quartz or hard-glass envelope containing a halogen (bromine or iodine) to avoid internal blackening of the envelope due to the deposition of tungsten from the filament.

tungsten lighting Lighting with incandescent lamps having tungsten filaments, including tungsten-halogen types.

turnkey Supplying a complete system, self-contained and ready to use; e.g. a computer with both the hardware and the software required for an application.

turntable (1) A constant-speed rotating platform to carry a disc, typically a gramophone record, during reproduction. (2) A mechanism fitted to an animation table or camera to produce *rotations*.

135

turret   A rotatable mount on a camera or projector allowing any one of two or more lenses to be brought into position on the optical axis as required; see *Figure T.4.*

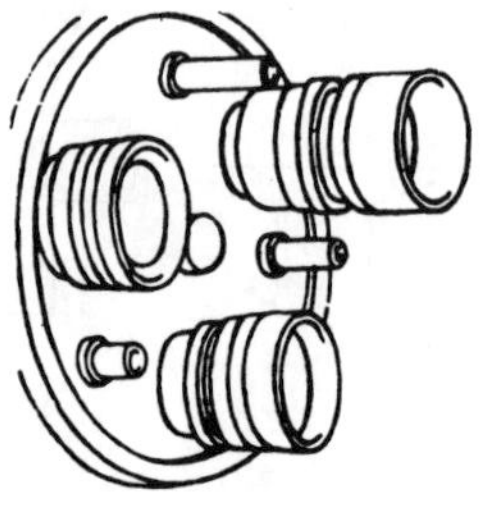

Figure T.4 Turret

TV   Television.

TVRO   Television Receive Only, equipment for reception and display without transmission.

tweeter   A loudspeaker designed to handle only the high frequencies of the audio spectrum.

twinkle   In slide projection, a rapid image intensity fluctuation effect.

twinning stand   A stand which supports one automatic slide projector on top of another, which actually bears its weight.

twitter   Defect of video picture showing flicker at the horizontal borders of objects.

two-duration control   The control of two slide-projectors to give fast and slow dissolve changes by means of cue tones of 1000 Hz but of two different durations, typically 50 ms and 450 ms.

TWT   Travelling Wave Tube amplifier.

TX   Transmission.

U

UART   Universal Asynchronous Receiver-Transmitter. An integrated circuit for serial/parallel signal conversion and vice versa.

U-format   Videocassette format using ¾ inch tape compatible with the U-matic system.

UHF   Ultra High Frequency of radio waves. 300 MHz to 3000 MHz.

UJT   Unijunction Transistor.

ULA   Uncommitted Logic Array.

ultrasonic   Sound at frequencies above the upper audible limit for humans.

ultrasonic cleaner   A machine for cleaning motion picture film by passing it through a solvent agitated at high (ultrasonic) frequencies.

ultra-violet   Invisible radiation at wavelengths shorter than the violet end of the visible spectrum, from 390 nm down. Generally detectable by photographic means, fluorescence, or by suitable photo-diodes or photo-transistors.

unblooped   A splice in a photographic sound track which has not been *blooped* and which may therefore produce a noise on reproduction.

U-matic   Video cassette system using ¾ inch tape. (Trade name).

unbalanced   Referring to a two-wire circuit one side of which is operating at or near ground potential. See *ground loop*.

under-cranking   Running a camera at less than normal speed so as to speed up an action on projection.

under-damped   *damping* in which a steady state is reached only after one or more cycles of oscillation; see *Figure D.1*.

unidirectional   Sensitive in one direction only.

unisette   Type of audio ¼ inch tape cassette developed by BASF for professional use.

Unix   A computer operating system.

unmod   Unmodulated, a photographic sound track on which no signal has been recorded.

unsqueezed   A motion picture print in which the compressed image of an *anamorphic* negative has been corrected for normal projection.

up rate   The speed of the up-coming lamp in a slide projector dissolve change system.

up stage   Performing area furthest from the audience or camera.

USART   Universal Synchronous/Asynchronous Receiver-Transmitter. See *UART*.

UV   Ultra-Violet.

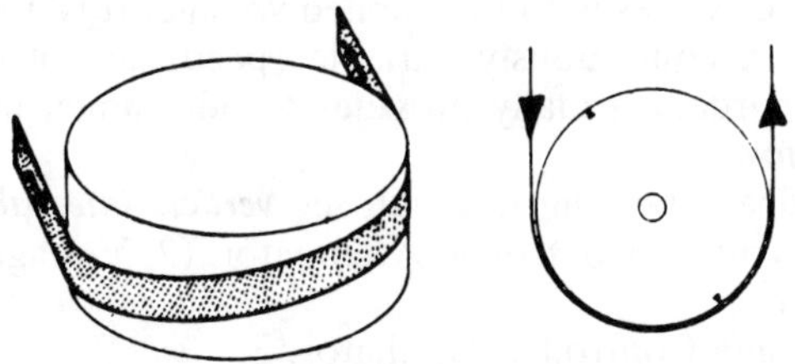

Figure U.1 U-wrap. The tape in the helical-scan path is in contact with the recording head drum over 180°

UV eraser   A device containing a short wave ultra-violet light source. Used to erase *EPROM* so that they can be reprogrammed.

U-wrap   Tape path in a helical-scan videotape system giving 180° contact on the drum, thus requiring the use of two or more rotating heads; see *Figure U.1*.

V

V   Volt. The unit of electrical pressure.

V24   A CCITT list of interchange circuits for connecting a terminal to a *modem*. See *RS232*.

VA   Volt-Amperes. A measurement of AC power. The product of voltage and current in an AC circuit, without regard to phase angle. Due to inductance and/or capacitance altering the voltage/current phase angle, VA may not be the same as watts.

valve rollers   See *fire rollers*.

VAR   Volt Ampere Reactive.

variable area track   Photographic sound track in which the image width varies according to the sound modulation.

variable density track   Photographic sound track in which the density of the image varies according to the sound modulation.

variable-frequency control   A control system for controlling two slide projectors using a continuous tone of varying frequency. Also called *FM control*.

variable transformer   A transformer with a smoothly variable output voltage.

variables   In computer practice, variable names used to refer to items of data; the computer assigns storage area for each name and its associated data. There can be three types, *integer*, *real* and *string*.

Variac   A continuously variable transformer. (Trade name).

varicap diode   A special diode having the property that its capacitance varies with the applied voltage. It is used in tuners.

varispeed   A continuously variable speed control on magnetic audio recorders; usually provides a wide range, but may be a *pitch control*.

VBI   Vertical Blanking Interval, see *vertical interval*.

VCA   (1) Voltage Controlled Attenuator. (2) Voltage Controlled Amplifier.

VCO   Voltage Controlled Oscillator.

VCPS   Video Copyright Protection Society (UK).

VCR   Video Cassette Recorder. (Trade name).

VDA   (1) Video Distribution Amplifier. (2) Video Dealers' Association (UK).

VDR Voltage-Dependent Resistor.

VDT Visual Display Terminal used for communicating with computers.

VDU *visual display unit.*

vector (1) A representation of magnitude and direction. (2) A line connecting defined points on a cathode ray tube. (3) A pointer used to allow access to relocatable program or data.

vector graphics A method of presenting graphical data in which the display is composed of *vectors* drawn directly from point to point on the cathode ray tube instead of using a *raster*, thus avoiding *aliasing*. (= *vector mode display*).

vectorscope In television, a special oscilloscope used for monitoring colour bars.

vertical interval Field blanking period between pictures. Cuts taking place in the vertical interval are not visible on the picture.

VHD Video High Density, grooveless capacitance videodisc system. (Trade name).

VHF Very High Frequency, of radio waves. 30 MHz to 300 MHz.

VHS Video Home System, a videocassette system, using ½ inch tape. (Trade name).

VHS-C Compact video cassette system using the *VHS* format.

VIBGYOR A mnemonic for the order of colours in the visible spectrum, i.e. Violet, Indigo, Blue, Green, Yellow, Orange, and Red. Outside this range are Ultra-violet and Infra-red.

video (1) Electronic means of recording, storing and reproducing visual images. (2) Incorrectly used for a home video recorder, or a *videogram*.

video cassette A cassette containing video recording tape with separate supply and take-up spools. In the recorder or player, a loop of tape is drawn into the machine and returned to the cassette. (In contrast to *cartridge* or *reel-to-reel*).

video conferencing The use of bi-directional vision and sound links to permit a conference between two or more participants at different locations.

videodisc A video and audio recording on a flat rotating medium, usually using digital storage, by optical (reflected laser) or mechanical (capacitive) means.

video dub Recording or re-recording the picture and sync control tracks of a videotape without disturbing the existing audio signal.

videogram The programme content of a video cassette or video-disc.

videographics Direct manual creation of graphics on a *VDU* for videotape recording, not by computer generation.

videography Telecommunications involving selection and display of textual and pictorial information on a *VDU* (= *videotex*); *teletext* is broadcast videography.

videophone Still picture transmission over conventional telephone lines.

video signal The signal in its basic form, not modulated on to an RF carrier. The video signal cannot be fed directly to the aerial input of a domestic TV receiver.

videotape Magnetic tape specifically designed for use as a video recording medium.

videotex General term for an information service of selected alpha-numeric/graphic data, presented as a still-picture video display.

video theatre A place of entertainment, where moving pictures and sound are presented by projection television.

videowall Mosaic assembly, usually in a rectangular block, of a number of individual video screens, displaying multiple, individual or composite images from *VTR*, *videodisc* or computer sources.

Vidicon TV camera tube employing a photo-conductive image detector. (Trade name).

Viditel Dutch viewdata service.

Viewdata A *videotex* data service distributed by (telephone) cable, and accessed by direct dialling.

viewer Optical device for inspecting the picture image on a film or slide.

viewfinder An optical or video system allowing a cameraman to see the scene embraced by the camera lens.

vignetting Gradual shading of the edges surrounding a picture area.

vinegar syndrome Decompostion of acetate-base film, producing a characteristic odour.

VIPS Visual Information Processing System, trade name for a *interactive* computer/videodisc system by Sony.

virtual earth A type of amplifier presenting a zero input impedance. Current signals may be combined at the amplifier's input without interaction.

viscous process Photographic processing in which the chemicals are applied as layers of viscous solutions.

visual display unit A display device containing a cathode ray tube screen, on which information, often from a computer, can be shown. A keyboard is usually incorporated to select the required information.

VITC Vertical Interval Time Code, recorded between frames for video editing.

VITS Vertical Insertion Test Signals.

VLF   Very Low Frequency of radio waves. 3 kHz to 30 kHz.

VLP   Very Long Play, a videodisc system. (Trade name).

VLSI   Very Large Scale Integration.

VMOS   Vertical (or V-shaped) Metal-Oxide-Silicon.

VO   *voice over.*

vocoder   A device which breaks up speech into fixed frequency bands and extracts slowly varying signals representing the energy in each band. These extracted signals can be used to modulate other audio signals, such as instrumental sounds.

voice over   Narrative recording on a programme, heard without the speaker being seen.

voice track   A recording track reserved for narration or dialogue as opposed to music or effects.

volatile memory   Data memory where the data is lost if power fails or is switched off, e.g. semi-conductor memory.

voltage controlled amplifier   An amplifier in which the gain is proportional to a control voltage. It can serve as an automatic volume control.

voltage controlled oscillator   An oscillator in which the frequency is proportional to a control voltage. It can serve as an automatic frequency control.

voltage follower   A circuit, usually employing an *op amp*, where the output voltage is maintained close to the input voltage in spite of driving a low impedance load.

volume   The intensity or magnitude of sound.

VOM   Volt-Ohm Meter, a multimeter.

vox pops   In TV programmes, snap interviews with random members of the public.

VT   *videotape.*

VTR   Video Tape Recorder, to record and replay television pictures and sound on magnetic tape.

VTVM   Vacuum Tube Volt Meter. A high input impedance voltmeter.

VU meter   Volume Unit meter. An audio level meter with defined ballistics and an average rectifier characteristic.

W

W   *watt.*

walk through   A full rehearsal including camera movements, but without actually filming or recording sound. Also called a *dry run*.

warble tone   Frequency modulated tone.

watt   Unit of electrical power.

waveform   A presentation of the varying amplitude of a signal in relation to time.

waveform monitor   Oscilloscope for displaying and measuring the waveform of electrical signals.

wavelength   Distance between successive corresponding points in sound or electromagnetic waves.

Wb   Weber.

weave   Periodic lateral unsteadiness of a projected film image.

weber   Unit of magnetic flux.

weighting   A correction factor applied to a measurement. In particular, standardised modification of the frequency response of measuring instruments for *noise, rumble* and *wow* so that the indications correspond to subjective effects.

wet printing   See *liquid gate*.

whip pan   A *pan* of such speed that the image becomes entirely indistinct. In animation, whip pans have the artwork reduced to blurred lines between the initial and final fields.

white noise   Noise containing energy distributed uniformly over the frequency spectrum.

widescreen   In general, pictures presented with an aspect ratio greater than 1.4:1.

wild shooting   Pictures recorded without synchronised sound.

wild track   Sound recorded without a simultaneous picture.

winchester disk   A hard disk system for storing large amounts of data with a reasonably fast access time, i.e. faster than a floppy disk but slower than semi-conductor memory.

windshield   An attachment to reduce wind noise and blasting in a microphone.

window   In computer graphics, a specified area through which a portion of the image can be viewed; see *Figure W.1*.

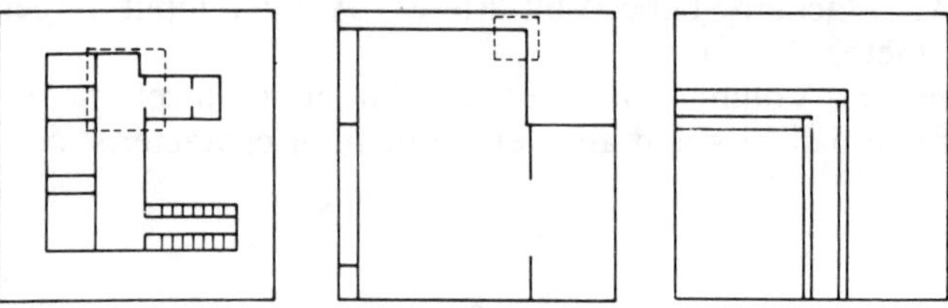

Figure W.1 Window. The selected area of a graphics display may be examined at different degrees of enlargement

windows   Computer technique for presenting information from more than one source on a *VDU*; some of the data appears in a window, which may be varied in size.

wings   Hidden spaces at the side of a stage, used for entrances and exits.

wipe   Visual transition in which one image is replaced by another at a boundary edge moving across the picture area.

wire frame   In computer graphics, a three-dimensional image displayed as a series of line segments outlining its surface, including hidden lines; see *Figure W.2*.

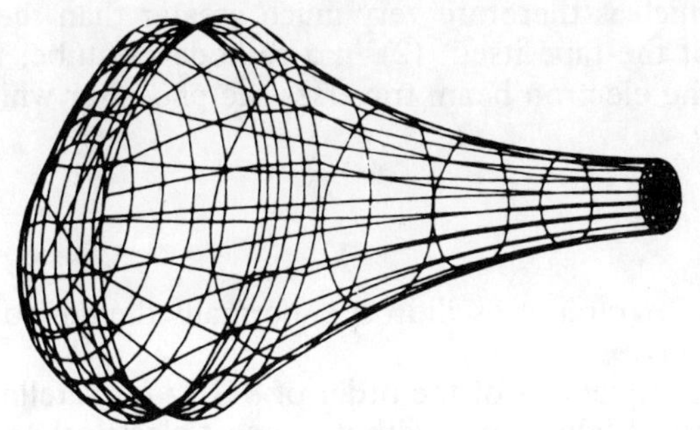

Figure W.2 Wire frame. A computer graphics image of
a three-dimensional form

woofer   A loudspeaker unit for reproducing low frequencies.

word (digital)   A set of bits, usually 8 or 16.

word wraparound   In word processing, if a word is too long to fit on a line, the whole word is moved to the start of the next line.

work print   In motion picture production, the assembled positive prints of scenes selected by the editor. Also called *cutting copy*.

WORM   Write Once, Read Many, in a data system which, once recorded, cannot be altered.

wow   Periodic variation in speed of a recording medium, occurring between 1 and 10 times per second.

wraparound   The positioning of an image in a computer graphics display so that it overlaps the specified boundaries, and is presented on the opposite side of the display surface; see *Figure W.3*.

write   To clock data into memory. To record data on to a magnetic medium.

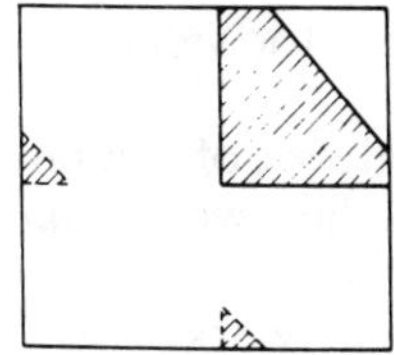

Figure W.3 Wraparound. Where the image in a computer graphics display overlaps the boundaries its extensions may be presented on the opposite sides

write-on slides Slides with a matt surface, used temporarily during production with hand-drawn information to indicate the final image.

write protected Guarded against accidental recording ('writing') on a magnetic medium, for example a *floppy disk* with its notch covered.

writing speed (1) In videotape systems, the speed at which the moving recording/replay head traverses the magnetic tape surface, which is therefore very much greater than the transport speed of the tape itself. (2) In a cathode ray tube, the rate at which the electron beam traverses the phosphor while writing.

X

X-**axis** In a graph or oscilloscope, normally the horizontal axis of the display.

X-band Frequencies of the order of 8 GHz for satellite use.

xenon lamp Light source with a compact electrical arc discharge within a quartz envelope containing the gas xenon at high pressure; widely used in motion picture film projectors.

xerography Method of producing images by the attraction of particles of dielectric powder to a charged surface, where they may be fixed on paper or plastic by heat; used in document copying and printing.

XLR connector A professional circular multi-pole connector of a locking type. Commonly used for audio frequency signals.

X-Y **plotter** A paper and ink plotting device where the pen can be driven on the *X* and *Y* *axes*. It may be used as a computer peripheral. Also known as a *data plotter*.

Y

Y The *luminance* component of a colour TV signal.

Y-**axis** In a graph or oscilloscope, normally the vertical axis of the display.

Y/C Separate *luminance* Y and *chroma* C signals for videotape *component* recording.

YIQ In the *NTSC* colour TV system, the complete set of component signals, comprising *luminance* Y and the two *colour difference signals*, I and Q.

YLE Oy Yleisradio, the Finnish state broadcast network.

yon plane   In a computer graphics display, the back *clipping* plane defining the limit of three-dimensional space furthest from the viewing plane; see *Figure Y.1*.

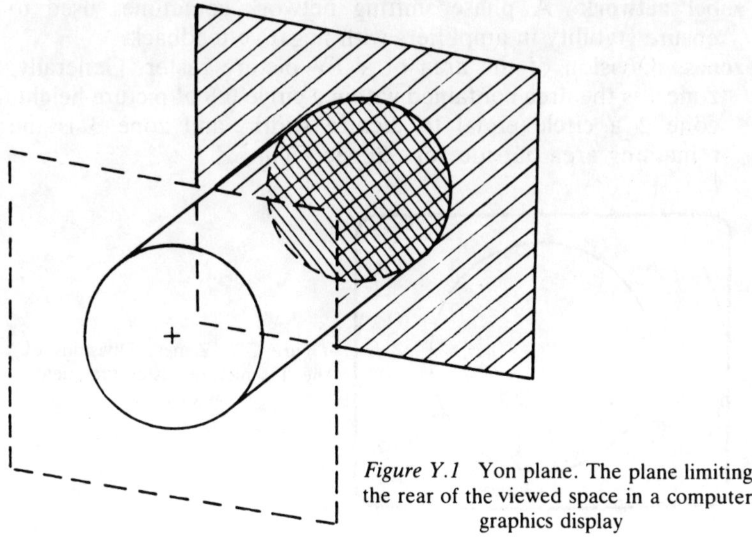

Figure Y.1   Yon plane. The plane limiting the rear of the viewed space in a computer graphics display

YUV   In the *PAL* colour TV system, the complete set of component signals, comprising *luminance* Y and the two *colour difference signals*, U (=B-Y) and V (=R-Y).

Z

z   Greenwich mean time.

Z-adjustment   Movement perpendicular to the plane of the X-Y axes. Specifically in an overhead projector, the lamp adjustment to reduce colour fringing when magnification and projection distance are changed.

Z-axis   Intensity modulation of an oscilloscope.

Z-clipping   In a computer graphics display, limiting the three dimensional space in depth by defining a *hither* and *yon* plane, both parallel to the view plane.

ZDF   Zweites Deutsches Fernsehen, the second West German state TV broadcast network.

zener diode   Semiconductor diode used as a voltage reference.

zero-crossing   The point on an AC waveform where the voltage crosses the x axis.

zero switch   The switch on a slide projector for detecting when the magazine is off-zero. Also called a homing switch.

ZIF   Zero Insertion Force, type of sockets commonly used with *EPROM* programmers.

zobel network   A phase shifting network sometimes used to ensure stability in amplifiers with negative feedback.

zones   Division of the area of a TV picture raster. Generally, zone 1 is the area contained within a circle 0.8 of picture height, zone 2 a circle equal to picture width, and zone 3 is the remaining area outside zone 2; see *Figure Z.1*.

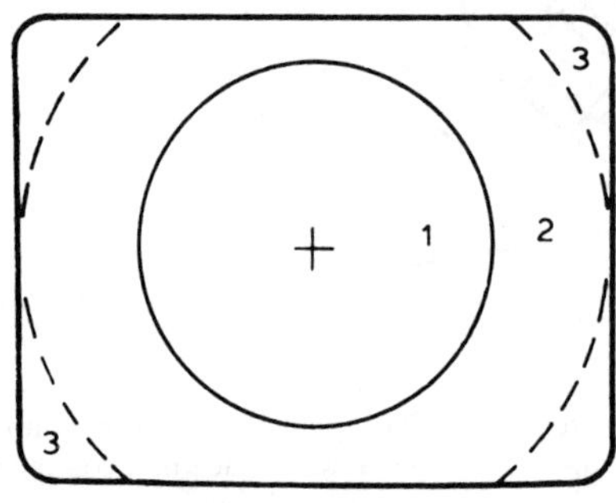

Figure Z.1 Zones. Divisions of the TV picture raster for quality assessment

zoom   (1) A visual effect as though the camera were moving rapidly towards or away from the subject. (2) In animation, moving the camera towards or away from the artwork to photograph larger or smaller fields.

zoom lens   An *objective* system for camera or projector whose magnification can be continuously varied while retaining a fixed focal plane.

zoom microphone   A microphone whose directional properties can be externally controlled to give a zoom effect.

Appendices

Appendix A   Image areas for motion picture films

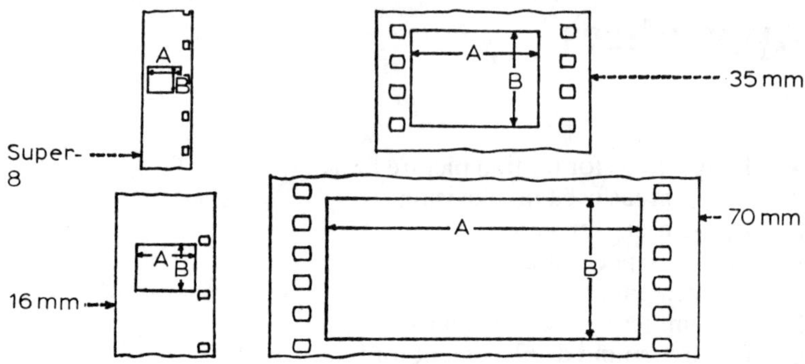

Film gauge	Camera Aperture (mm)		Maximum Projected Area (mm)	
	A	*B*	*A*	*B*
Super-8 mm	5.69	4.22	5.46	4.01
16 mm	10.05	7.42	9.65	7.26
Super-16 (Type W)	12.52	7.42	–	–
35 mm Academy	21.95	16.05	21.11	15.29
Wide Screen AR 1.85:1	21.95	12.00	21.11	11.41
Anamorphic	21.95	18.80	21.29	18.21
65 mm/70 mm	52.50	23.00	48.59	22.10

Appendix B Image areas used in television

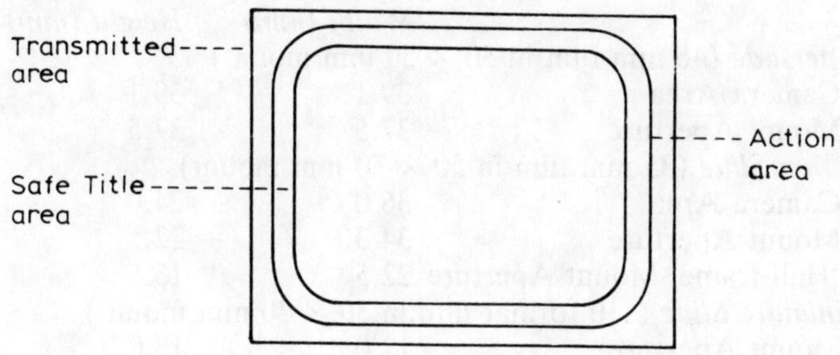

	Transmitted Area	*Action Area*	*Safe Title Area*	*Corner Radius*
16 mm film	9.35 × 7.00	8.4 × 6.3	7.5 × 5.6	1.5
35 mm film	20.12 × 15.10	18.1 × 13.6	16.1 × 12.1	3.2
50×50 mm slides	28.60 × 21.50	25.7 × 19.3	22.8 × 17.2	4.6

(Dimensions in millimetres)

The dimensions of the Action Area and Safe Title Area represent 90% and 80% respectively of the Transmitted Area.

Full details of cinematography standards are given in British Standard BS 5550, which is divided into the following sub-sections:

5550:1	8 mm & Super-8	5550:2	16 mm
5550:3	35 mm	5550:4	65 mm & 70 mm
5550:5	Common to several gauges		
5550:6	Television usage	5550:7	Production & Presentation

New American (ANSI/SMPTE) and international (ISO) standards are regularly published in the *Journal of the Society of Motion Picture and Television Engineers*. An annual summary of all current American, British and international standards appears in *Image Technology*, the journal of the British Kinematograph, Sound and Television Society.

Appendix C   Image areas for slides

	Width (mm)	Height (mm)
Superslide (46 mm film in 50 × 50 mm mount)		
Camera Area	39.1	39.1
Mount Aperture	37.5	37.5
35 mm Slide (35 mm film in 50 × 50 mm mount)		
Camera Area	36.0	24.0
Mount Aperture	34.3	22.5
'Half-frame' Mount Aperture	22.5	15.5
Miniature Slide (110 format film in 30 × 30 mm mount)		
Mount Aperture	17.0	13.0
35 mm Film Strip		
Camera Area	24.0	18.0
Mount Aperture	22.35	16.75

Appendix D   Picture aspect ratios

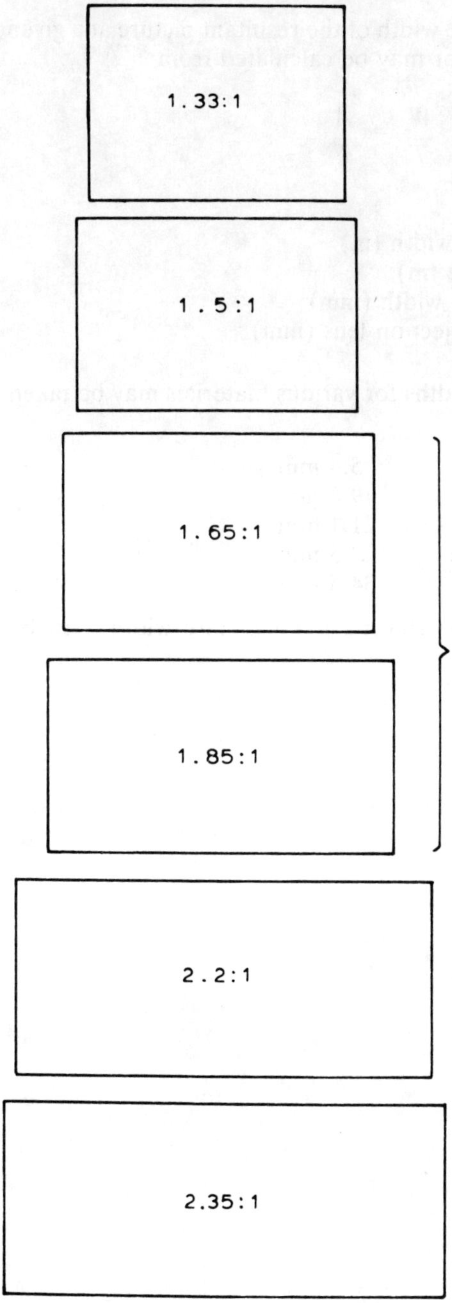

1.33:1	Television Super-8 film 16 mm film Film-strip
1.5:1	Slide projection
1.65:1	35 mm film Wide-screen presentation
1.85:1	
2.2:1	70 mm film
2.35:1	35 mm film Anamorphic projection

Appendix E Projected image size

For normal projection, the width of the resultant picture at a given distance from the projector may be calculated from:

$$\frac{W}{T} = \frac{A}{F}$$

where
  W = projected picture width (m)
  T = projection distance (m)
  A = projector aperture width (mm)
  F = focal length of projection lens (mm)

The projector aperture widths for various materials may be taken as

Super-8 film	5.4 mm
16 mm film	9.6 mm
35 mm film	21.1 mm
Film strip and half-slide	22.5 mm
Standard 50 mm slide	34.5 mm

For anamorphic projection, the resultant picture width must be doubled.

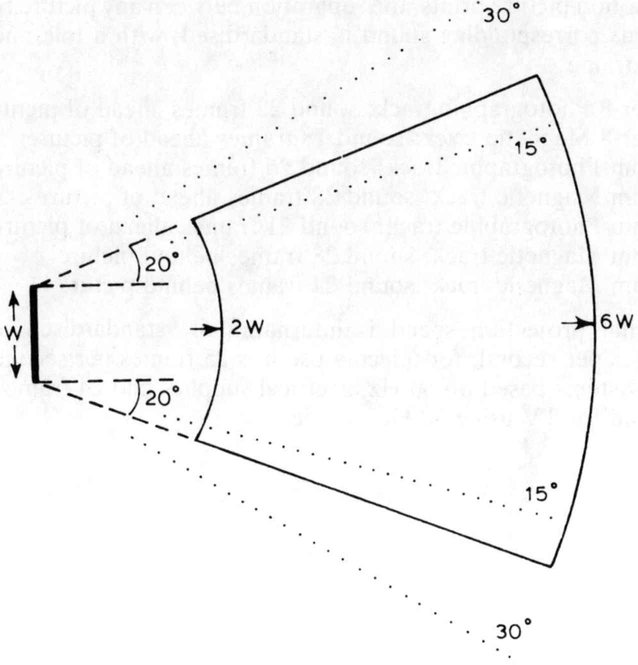

In general, the seating lay-out should have the front row not closer to the screen than two screen-widths, with the back row not more than six screen-widths away. Seats should preferably be kept within a wedge area of about 20° from each side of the screen but this can be extended to 30° for non-directional matt white surfaces. For more directional screen materials it is desirable to keep within 15°.

Appendix G   Picture–sound displacement

In motion picture prints, the separation between any picture frame and its corresponding sound is standardised, with a tolerance of $\pm\frac{1}{2}$ frame.

Super-8 Photographic track: sound 22 frames ahead of picture
Super-8 Magnetic track: sound 18 frames ahead of picture
16 mm Photographic track: sound 26 frames ahead of picture
16 mm Magnetic track: sound 28 frames ahead of picture
35 mm Photographic track: sound 21 frames ahead of picture
35 mm Magnetic track: sound 28 frames behind picture
70 mm Magnetic track: sound 24 frames behind picture

Normal projection speed is internationally standardised at 24 frames per second; for telecine use it is 25 frames per second for TV systems based on 50 Hz electrical supplies and 24 frames per second for TV using 60 Hz supplies.

Appendix H   Impairment and quality scales

In video and audio reproduction, the subjective grading scales for impairment and quality are:

Degree of Impairment	Scale	Quality Rating
Imperceptible	5	Excellent
Barely perceptible	4	Good
Perceptible, slightly objectionable	3	Fair
Objectionable	2	Poor
Unacceptable	1	Bad

Appendix I   Videocassette formats

Helical-scan Systems to PAL standard

	U-matic	Betamax	VHS	VKR 8500
Originator	Sony	Sony	JVC/Matsushita	Philips
Tape width, mm	19	12.5	12.5	8
Tape speed, cm/sec	9.5	1.87	2.34	2.0
Video writing speed, cm/sec	854	583	485	310
Head drum diameter, mm	110	74	62	40
Video heads	Two	Two	Two	Three
Head gap azimuth differences	0°	±7°	±6°	±10°
Cassette size, mm	221 × 140	156 × 96	189 × 104	140 × 100
Cassette thickness, mm	32	25	25	18
Cassette weight, g	435	210	280	170

Appendix J   Television single line waveform

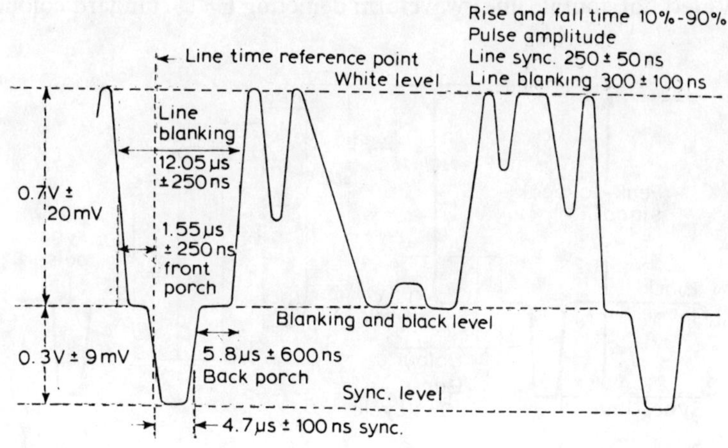

The waveform of a typical line showing synchronizing signals for a
625-line 50-field television system, with acknowledgement to the
IBA Technical Reference Book. In a 525-line 60-field system the
general form is similar but the pulse durations measured at
half-amplitude points are different.

	625/50	525/60
Line blanking	12.05 µs	11.11 µs
Front porch	1.55 µs	1.59 µs
Sync pulse	4.7 µs	4.76 µs
Back porch	5.8 µs	4.76 µs

Appendix K   Television colour bars and vectorscope display

Video horizontal (line) waveform depicting EBU standard colour

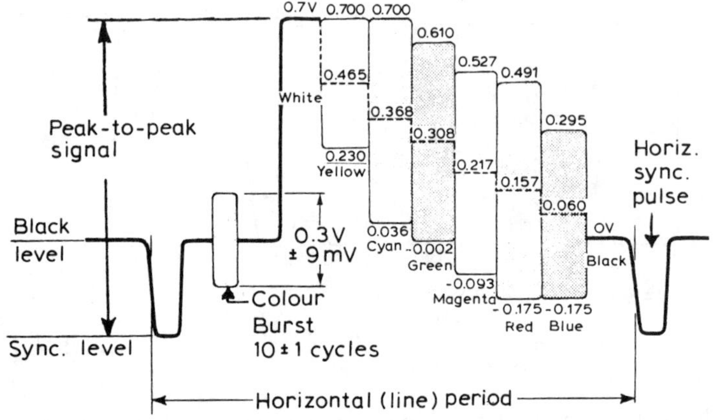

Vectorscope display of encoded colour bars on PAL system

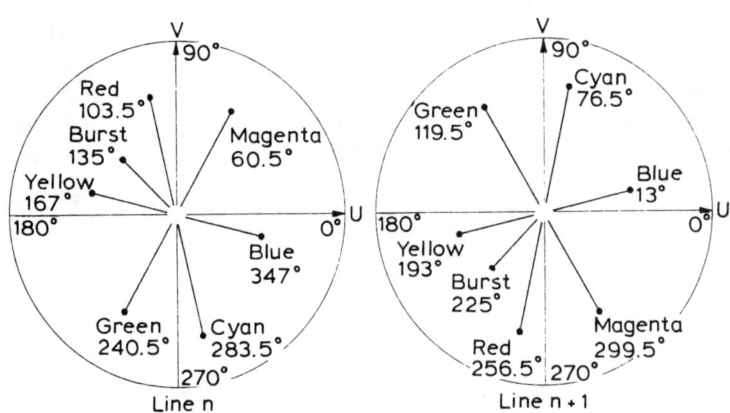

Appendix L   Prefixes and symbols for numerical factors

exa	E	10^{18}	atto	a	10^{-18}
peta	P	10^{15}	femto	f	10^{-15}
tera	T	10^{12}	pico	p	10^{-12}
giga	G	10^{9}	nano	n	10^{-9}
mega	M	10^{6}	micro	μ (mu)	10^{-6}
kilo	k	10^{3}	milli	m	10^{-3}
hecto	h	10^{2}	centi	c	10^{-2}
deca	da	10	deci	d	10^{-1}

Appendix M   The electro-magnetic spectrum

Band	*Frequency*	*Wavelength*	
VLF	<30 kHz	>10 km	Long wave radio (AM)
LF	30–300 kHz	10–1 km	
MF	300–3000 kHz	1000–100 m	Medium wave radio (AM)
HF	3–30 MHz	100–10 m	
VHF	30–300 MHz	10–1 m	Short wave radio (FM)
UHF	300–3000 MHz	1–0.1 m	Television
SHF	3–30 GHz	100–10 mm	Satellite TV
EHF	30–300 GHz	10–1 mm	Microwave
		1000–0.7 µm	Infra-red
		700–400 nm	Visible light
		400–0.1 nm	Ultra-violet
		100–0.1 pm	X-rays
		<100 fm	Gamma rays

Frequency (Hz) × Wavelength (m) = c, the velocity of light
$$= 2.988 \times 10^8 \text{ metres per second}$$